Django in Production

Expert tips, strategies, and essential frameworks for writing
scalable and maintainable code in Django

Arghya Saha

Django in Production

Group Product Manager: Rohit Rajkumar

Publishing Product Manager: Jane Dsouza

Book Project Manager: Aishwarya Mohan

Senior Editor: Rashi Dubey

Technical Editor: K Bimala Singha

Copy Editor: Safis Editing

Indexer: Hemangini Bari

Production Designer: Alishon Mendonca

DevRel Marketing Coordinators: Nivedita Pandey and Anamika Singh

First published: April 2024

Production reference: 1070324

Published by Packt Publishing Ltd.

Grosvenor House

11 St Paul's Square

Birmingham

B3 1RB, UK

ISBN 978-1-80461-048-0

www.packtpub.com

To my "Maa," Anima Saha, and to the memory of my "Baba," for their constant support in helping me reach where I am today. To my partner, Parul, for being a constant inspiration and motivation.

– Arghya Saha

Contributors

About the author

Arghya (argo) Saha, is a software developer with 8+ years of experience and has been working with Django since 2015. Apart from Django, he is proficient in JavaScript, ReactJS, Node.js, Postgres, AWS, and several other technologies. He has worked with multiple start-ups, such as *Postman* and *HealthifyMe*, among others, to build applications at scale. He currently works at *Abnormal Security* as a senior Site Reliability Engineer to explore his passion in the infrastructure domain.

In his spare time, he writes tech blogs. He is also an adventurous person who has done multiple Himalayan treks and is an endurance athlete with multiple marathons and triathlons under his belt.

This book is dedicated to my Maa and memories of my Baba. I would like to thank my partner, Parul, and all my friends who supported me throughout this journey. I am grateful to Ganesh, Rahul, and everyone who helped me with my book.

About the reviewers

Abdul-Rahman Mustafa Saber Abdul-Aziz is an Egyptian Python backend developer, senior AI instructor, capstone projects lead, and IBM consultant. He graduated from Assiut University in Upper Egypt. He believes that to be great, you must take great responsibility.

I want to thank my family, my friends, and my future wife, who I haven't met yet. I love you so much. I want to thank my father once again; everything I am now is thanks to him. This is the second time that my name has appeared on the Contributors' list of a book. I want to thank Dr. Rania Hafez, Dr. Huda Goda, Dr. Abdul-Rahman Eliwa, and Mr. Ehab, who always pushed me toward progress, and my close friends, Muhammad Adly, Ahmed Fouad, and many others. Thank you all for everything you gave me; you are my family.

Ruben Atinho, a software engineer specializing in backend engineering, explores the vast realms of technology. Proficient in Python, Django, and PostgreSQL, and experienced with Golang, he combines technical expertise with a passion for reading open source code. Beyond the code base, Ruben finds fulfillment in the harmonies of music, insightful articles, captivating books, and the imaginative narratives of anime.

Md Enamul Hasan is a seasoned full stack Python developer with over a decade of professional experience in software and web application development. Based in New York, United States, Enamul is renowned for his expertise in Python, Django, React, and AWS. He has successfully led the development of various software products, including ERP, e-commerce, and data-driven applications.

In addition to his technical proficiency, Enamul has excelled in leadership roles, contributing significantly to project success. He actively shares his knowledge, participating in forums and conferences and showcasing a passion for problem-solving in the ever-evolving software development landscape.

Enamul extends gratitude to the author for the opportunity to review *Django in Production*. His extensive experience in Python Django development adds depth to the review, highlighting the book's value in bridging theory and practice in the dynamic tech industry.

Table of Contents

3

Serializing Data with DRF   69

4

Part 2 – Using the Advanced Concepts of Django

6

7

8

9

Writing Tests in Django   171

10

Exploring Conventions in Django   197

Part 3 – Dockerizing and Setting Up a CI Pipeline for Django Application

11

12

Part 4 – Deploying and Monitoring Django Applications in Production

13

Deploying Django in AWS 253

14

Monitoring Django Application 285

Preface

Hey there! As the name suggests, Django in Production is a book to help developers put their application code into production. In today's world, coding has become a profession that people get into after joining a 3–6 month boot camp. With the start-up boom, most of these developers are able to land a job after their boot camp course, since they are able to write code and hack any product together. However, a couple of months into the job, they want to learn about the best practices and understand all the aspects that senior developers in the industry know and perform, but most start-ups don't have many senior developers due to budget and talent scarcity. This book is going to give them a too long; didn't read (TLDR) version of software development best practices, which they need to know to get to the next level.

Who this book is for

This book is for any software developer who understands the basic concepts of Django but now needs some help putting their code to production using the right tools, or someone who does not have enough guidance to know how to do the work the right way. We are assuming you have a basic understanding of how to write code in Django and now want to improve your skills.

What this book covers

Chapter 1, Setting Up Django with DRF, covers the basic project setup of Django and **Django Rest Framework** (**DRF**). It will also help you to understand the fundamentals of APIs and how to design a REST API.

Chapter 2, Exploring Django ORM, Models, and Migrations, covers how to integrate Django with a database. Django ORM and migrations are powerful features; we learn about the core concepts and how to use them efficiently in this chapter.

Chapter 3, Serializing Data with DRF, explores the concept of serialization and how developers can use DRF serializers to write better application code.

Chapter 4, Exploring Django Admin and Management Commands, covers the core concepts of Django admin. This chapter covers all the best practices on how to use Django admin and create custom Django management commands.

Chapter 5, Mastering Django Authentication and Authorization, covers the key concepts of authentication and authorization. Django provides authentication and authorization out of the box, and we will explain in detail how developers can use the built-in features of Django and DRF for authentication.

Chapter 6, Caching, Logging, and Throttling, covers all the concepts of caching and how to integrate Redis with Django for caching. Logging is a crucial component of any web application in production and, in this chapter, we will learn how to integrate logging into a Django application.

Chapter 7, Using Pagination, Django Signals, and Custom Middleware, covers all the advanced concepts of Django. Developers can use Django signals to write decoupled code. Django also gives the flexibility to create custom middleware that can help developers to improve features.

Chapter 8, Using Celery with Django, shows how to process asynchronous tasks for web applications. In this chapter, developers will get an understanding of how to integrate Celery into a Django project.

Chapter 9, Writing Tests in Django, covers the core concepts of writing test cases for Django. In this chapter, you will learn the best practices to follow while writing test cases and understand the importance of writing test cases for a project.

Chapter 10, Exploring Conventions in Django, shows all the best practices and conventions that are used while working with Django. This chapter covers a lot of concepts that are opinionated, and you are expected to read this chapter as an outline and pick/learn about concepts by using your own judgment.

Chapter 11, Dockerizing Django Applications, covers how to integrate Docker with a Django application.

Chapter 12, Working with Git and CI Pipelines Using Django, covers the concepts of version control and how to efficiently use Git in a Django project. In this chapter, you will learn how to integrate GitHub Actions to create a CI pipeline.

Chapter 13, Deploying Django in AWS, covers how to deploy Django applications in production using different AWS services. In this chapter, you will learn how to deploy and scale the Django application in production.

Chapter 14, Monitoring Django Applications, covers how to monitor Django applications in production. There are different types of monitoring needed in production, such as error monitoring, application performance monitoring, uptime monitoring, and so on. In this chapter, you will learn how to integrate different tools available on the market to monitor Django applications.

To get the most out of this book

You will need to have a basic understanding of Django and should be comfortable in writing basic Django application code. In this book, we will learn about many of the core concepts of Django and you need to be able to follow those code examples. We will introduce a lot of third-party tools/platforms that may be paid/ free, and you are expected to create an account on these platforms and integrate them into the Django project.

Software/hardware covered in the book	Operating system requirements
Python 3.10 and above	Windows, macOS, or Linux
Django 4.x, Django 5.0 and above	
Python packages such as `celery`, `django-fsm`, `factory-boy`, `freezetime`, `django-json-widget`, `rest_framework`	
Docker	
Amazon Web Services (AWS), ElephantSQL, Neon (`https://neon.tech`), Redis	
Tools such as Rollbar, NewRelic, Better Uptime.	

If you are using the digital version of this book, we advise you to type the code yourself or access the code from the book's GitHub repository (a link is available in the next section). Doing so will help you avoid any potential errors related to the copying and pasting of code.

Download the example code files

You can download the example code files for this book from GitHub at `https://github.com/PacktPublishing/Django-in-Production`. If there's an update to the code, it will be updated in the GitHub repository.

We also have other code bundles from our rich catalog of books and videos available at `https://github.com/PacktPublishing/`. Check them out!

Conventions used

There are a number of text conventions used throughout this book.

`Code in text`: Indicates code words in text, database table names, folder names, filenames, file extensions, pathnames, dummy URLs, user input, and Twitter handles. Here is an example: "Since we have specified the `DemoViewVersion` class, this view would only allow the `v1`, `v2`, and `v3` versions in the URL path; any other version in the path would get a `404` response."

A block of code is set as follows:

```
urlpatterns = [
    ...
    path('apiview-class/', views.DemoAPIView.as_view())
]
```

When we wish to draw your attention to a particular part of a code block, the relevant lines or items are set in bold:

```
urlpatterns = [
    path('hello-world/', views.hello_world),
    path('demo-version/', views.demo_version),
    path('custom-version/', views.DemoView.as_view()),
    path('another-custom-version/', views.AnotherView.as_view())
]
```

Any command-line input or output is written as follows:

```
celery --app=config beat --loglevel=INFO
```

Bold: Indicates a new term, an important word, or words that you see onscreen. For instance, words in menus or dialog boxes appear in **bold**. Here is an example: "Click on the **Create New Instance** button."

> **Tips or important notes**
> Appear like this.

Get in touch

Feedback from our readers is always welcome.

General feedback: If you have questions about any aspect of this book, email us at customercare@ packtpub.com and mention the book title in the subject of your message.

Errata: Although we have taken every care to ensure the accuracy of our content, mistakes do happen. If you have found a mistake in this book, we would be grateful if you would report this to us. Please visit www.packtpub.com/support/errata and fill in the form.

Piracy: If you come across any illegal copies of our works in any form on the internet, we would be grateful if you would provide us with the location address or website name. Please contact us at copyright@packt.com with a link to the material.

If you are interested in becoming an author: If there is a topic that you have expertise in and you are interested in either writing or contributing to a book, please visit authors.packtpub.com.

Share Your Thoughts

Once you've read *Django in Production*, we'd love to hear your thoughts! Scan the QR code below to go straight to the Amazon review page for this book and share your feedback.

https://packt.link/r/1804610488

Your review is important to us and the tech community and will help us make sure we're delivering excellent quality content.

Download a free PDF copy of this book

Thanks for purchasing this book!

Do you like to read on the go but are unable to carry your print books everywhere?

Is your eBook purchase not compatible with the device of your choice?

Don't worry, now with every Packt book you get a DRM-free PDF version of that book at no cost.

Read anywhere, any place, on any device. Search, copy, and paste code from your favorite technical books directly into your application.

The perks don't stop there, you can get exclusive access to discounts, newsletters, and great free content in your inbox daily

Follow these simple steps to get the benefits:

1. Scan the QR code or visit the link below

https://packt.link/free-ebook/9781804610480

2. Submit your proof of purchase
3. That's it! We'll send your free PDF and other benefits to your email directly

Part 1 –
Using Django and DRF to
Build Modern Web Application

In the first part of the book, we will get an overview of how to use Django and **Django Rest Framework (DRF)** to create a modern web application. We can expect to learn all the concepts related to Django ORM and DRF serializers, which are crucial to building any modern web application. Django Admin and Authentication are one of the most widely appreciated features of Django. We will learn all the best practices that a developer should know about before using Django and DRF in production.

This part has the following chapters:

- *Chapter 1, Setting Up Django with DRF*

- *Chapter 2, Exploring Django ORM, Models, and Migrations*

- *Chapter 3, Serializing Data with DRF*

- *Chapter 4, Exploring Django Admin and Management Commands*

- *Chapter 5, Mastering Django Authentication and Authorization*

1
Setting Up Django with DRF

In 2003, the **Django** project was started by developers *Adrian Holovaty* and *Simon Willison* from *World Online*, a newspaper web operation company, and was open sourced and first released in the summer of 2005. When Django was first built, the world was still using dial-up modem internet connections, mobile devices were still not popular, smartphones didn't see the daylight, and people would access web pages through their desktops and laptops. Django was the perfect framework that had all the features needed to build a web application for that age.

Over the last two decades, technology has evolved drastically:

- We have moved from dial-up internet connections to 4G/5G internet connections
- 55% of the world's internet traffic came from mobile devices in 2022 (`https://radar.cloudflare.com/`)

In this book, we shall see how to build a modern web application using Django and deep dive into the core concepts that a developer should know to create a scalable web application for startups. A developer building a product for a startup is expected to be more than just a regular developer writing code in Django; they are expected to develop their code, write tests for the business logic, deploy their applications to the web, and finally keep monitoring the service they have deployed. Here, we will learn how easy it is to develop web applications with Django and the best practices that developers in the industry follow, especially in startups, to make their development cycle easier and faster.

In this first chapter, we shall learn the basics of Django and how to set up a Django project and structure the project folders. Since we shall mostly work with RESTful APIs throughout this book, we will learn about the conventions of the REST API and the crux of setting up a Django project with **Django Rest Framework** (**DRF**) for creating RESTful APIs. We shall also focus on versioning APIs and how we can implement versioning using DRF. DRF gives us the flexibility to create both functional and class-based views; we shall learn about them in this chapter, along with their pros and cons.

We will cover the following topics:

- Why Django?
- Creating a "Hello World" web app using Django and DRF
- Creating RESTful endpoints with DRF
- Working with views using DRF
- Introducing API development tools

Technical requirements

In this chapter, we shall do the basic project setup and also write our first Hello World app. Though this book is for developers who already know how to write a basic web application, anyone with decent programming skills can pick up this book and learn how to create a scalable Django web application. The following are the skill sets that you should possess to follow this chapter:

- Good Python programming knowledge and familiarity with packages and writing loops, conditional statements, functions, and classes in Python.
- A basic understanding of how web applications work and have written some form of API or web app before.
- Even though we shall try to cover most of the concepts from scratch, having basic knowledge of Django would be great. The Django Girls tutorial is a good resource to learn the basics: `https://tutorial.djangogirls.org/en/`.

You can find the code for this chapter in this book's GitHub repository: `https://github.com/PacktPublishing/Django-in-Production/tree/main/Chapter01`.

> **Important note**
>
> If you have any doubts about any of the topics mentioned in this or other chapters, feel free to create GitHub issues that specify all the relevant information (`https://github.com/PacktPublishing/Django-in-Production/issues`) or join our *Django in Production* Discord channel and ask us. Here is the invite link for the Discord server, where you can reach me directly: `https://discord.gg/FCrGUfmDyP`.

Why Django?

Django is a web framework based around Python, one of the most popular and easy-to-learn coding languages out there. Since Python is the go-to language for data science and artificial intelligence/machine learning, developers can easily learn Django without having to learn an additional language for building web applications.

Django's tagline, "*Django – The web framework for perfectionists with deadlines,*" proves its commitment to faster and more efficient development, further emphasized by its batteries-included principle that all the basic and widely used functionalities come out of the box with the framework rather than us having to install additional packages. This gives Django an additional advantage over other frameworks, such as Flask.

What is available with Django?

Django has evolved in the last decade and is currently in version 5.x, which has some promising new features, such as asynchronous support. However, the core modules of Django are still the same, with the same principles. When a new developer wants to learn Django, an organization wants to pick Django for their new project, or a startup with limited resources is looking to pick the perfect framework for their tech stack, they want to know why they should learn about Django. To answer this question, we shall learn more about the features of Django.

Let's look at the salient features of Django that the framework provides out of the box:

- In any organization, speed of execution is very important for the success of a product. Django was designed to help developers take applications from the concept phase to the product phase at blazing speed.

- Django takes care of user authentication, content administration, site maps, RSS feeds, and many more fundamental web tasks that developers look for in any framework.

- Security is a serious concern for any organization and Django helps developers avoid common security pitfalls.

- Websites such as Mozilla, Instagram, Disqus, and Pinterest all are built using Django, which makes Django a battle-tested framework that scales.

- Django's versatile framework can be used for different purposes, from content management systems to social networks to scientific computing platforms.

But the question of Django still being relevant is very subjective. Ultimately, it depends upon the use case. We know Django is a good web development framework, however, because more than 55% of the world's internet traffic comes from mobile devices using Android or IOS apps, you may be wondering whether Django is relevant for building features for mobile users? Django as a standalone framework might not be sufficient for today's modern web development where more and more organizations are moving towards API first development, but when integrated with frameworks like Django Rest Framework (DRF), Tastypie, etc, Django becomes the go-to framework for developers.

For start-ups with limited time and resources, it becomes even more crucial to choose a framework where they don't have to build every feature from the ground up, but rather leverage the framework to do most of the heavy lifting.

Let's quickly look a little more at the framework principle that Django uses: the MVT framework.

What is the MVT framework?

Most of us have heard of **MVC frameworks** (*Model-View-Controller*), which represent a paradigm of modern web frameworks where we have the following:

- **Model** represents the data and business logic layer
- **View** represents how the data is presented to the user in the UI/design layout
- **Controller** updates the model and/or view based on the user's input

Django considers the standard names debatable, hence why it has its own interpretation of MVC. Here, we have the following:

- **View** represents *which* data is shown to the end user and not how the data is represented
- **Template** represents *how* the data is represented to the end user
- **Model** represents the data layer

That's why Django follows the **MVT framework** (*Model-View-Template*). But now, the question is, what is the controller in Django? The framework itself is the controller since it handles the whole routing logic using its built-in features.

> **Important note**
> You don't need to deep dive into MVT concepts since this concept becomes muscle memory as you write more code in Django.

MVT is a concept where we use templates, but in today's world, most of the products are built for multiple domains such as mobile, IoT, and SaaS platforms. To build products for all these domains, the developer ecosystem has also evolved; now, organizations are moving toward an **API-first** development approach (`https://blog.postman.com/what-is-an-api-first-company/`). This means that APIs are "first-class citizens"; every feature in the product is built with an API-first model, which helps in creating a better client (mobile apps, frontend applications, and so on) and server integration. It involves establishing a contract between the client and the server so that each team can work in parallel without much dependency. Once both teams finish their work, the integration and development cycle of a product becomes much faster with a better developer experience.

The growing use case of mobile device means it is important to build platform-agnostic backend APIs that can be consumed by any client, Android app, iOS app, browser frameworks, and so on. Is Django, as an individual MVT framework, able to serve all these needs? No. The amount of additional effort required to use the out-of-the-box features of Django for creating APIs is similar to reinventing the wheel. That's why most organizations use Django's REST framework, along with Django, to create APIs.

> **Important note**
>
> In this book, instead of focusing on templates and standalone web applications built with Django, we shall focus on creating APIs using Django with DRF. For information on getting started with just Django, one of my favorite resources is the Django Girls tutorial: `https://tutorial.djangogirls.org/en/`.

Now that we have seen what the MVT framework is and how Django is an MVT framework, let's create a basic Hello World web application using Django and set up our project structure and development environment.

Creating a "Hello World" web app with Django and DRF

As mentioned previously, Django is a Python-based web framework, so we need to write the code using the Python programming language. If you are already using Linux or macOS-based systems, then Python comes preinstalled. However, for Windows systems, you have to install it by following the instructions on the official Python website: `https://www.python.org/downloads/`.

We shall also use **virtualenv** as our preferred tool to manage different environments for multiple projects, allowing us to create isolated Python environments.

> **Important note**
>
> We are not going to deep dive into `virtualenv` since we expect you to know how and why we use `virtualenv` for different projects. You can find details about `virtualenv` on its official website: `https://virtualenv.pypa.io/en/latest/index.html`.

First, let's create a virtual environment with the latest Python version (preferably >3.12.0). The following commands will work for Linux/Unix/macOS; for Windows, please check the next section:

```
pip install virtualenv
virtualenv -p python3 v_env
source /path to v_env/v_env/bin/activate
```

Now, we will break down what the code means:

- `pip install virtualenv` installs `virtualenv` on the system. `pip` is the built-in package manager that comes with Python and is already preinstalled on Mac and most Linux environments.

- `virtualenv -p python3 v_env` creates a new virtual environment with the name `v_env` (this is just the name we have given to our virtual environment; you can give another relevant name). The `-p python3` flag is used to tell us which interpreter should be used to create the virtual environment.

- `source /path to v_env/v_env/bin/activate` executes the `activate` script, which loads the virtual Python interpreter as our default Python interpreter in the shell.

Now that the Python virtual environment has been set up, we shall focus on managing the package dependency. To install the latest release of Django, run the following command:

```
pip install Django==5.0.2
```

For Windows systems, download Python 3.12 or higher from `https://www.python.org/downloads/windows/` and install it by following the wizard. Remember to click the **Add python. exe to PATH** checkbox in the installation step.

To verify your Python installation, use the following command in the terminal:

```
C:\Users\argo\> python --version
Python 3.12.0
```

Once Python has been installed successfully, you can use the following command to set up a virtual environment and install Django:

```
py -m pip install --user virtualenv
py -m venv venv
.\<path to venv created>\venv\Scripts\activate
// to install Django
pip install Django==5.0.2
```

The explanation for the Windows-specific commands is the same as what we explained for Linux/MacOS systems.

> **Important note**
>
> We are not using poetry, PDM, pipenv, or any other dependency and package management tools to avoid overcomplicating the initial setup.
>
> Furthermore, we prefer to use a Docker environment to create more isolation and provide a better developer experience. We shall learn more about Docker in *Chapter 10*.

With the previous command, our local Python and Django development environments are ready. Now, it's time to create our basic Django project.

Creating our Django hello_world project

We all love the `django-admin` command and all the boilerplate code it gives us when we create a new project or application. However, when working on a larger project, the default project structure is not so helpful. This is because when we work with Django in production, we have many other moving parts that need to be incorporated into the project. Project structure and other utilities that are used with a project are always opinionated; what might work for you in your current project might not work in the next project you create a year down the line.

> **Important note**
>
> There are plenty of resources available on the internet that will suggest different project structures. One of my favorites is django-cookiecutter. It gives you a lot of tools integrated into the project and gives you a structure that you can follow, but it can be daunting for any new beginner to start since it integrates a lot of third-party tools that you might not use, along with a few configurations that you might not understand. But instead of worrying about that, you can just follow along with this book!

We shall create our own minimalistic project structure and have other tools integrated with our project in incremental steps. First, let's create our hello_world project with Django:

```
mkdir hello_world && cd hello_world
mkdir backend && cd backend
django-admin startproject config .
```

Here, we have created our project folder, hello_world, and then created a subfolder called backend inside of it. We are using the backend folder to keep all the Django-related code; we shall create more folders at the same level as the backend subfolder as we learn more about the CI/CD features and incorporate more tools into the project. Finally, we used the Django management command to create our project.

> **Important note**
>
> Note the . (dot), which we have appended to the startproject command; this tells the Django management command to create the project in the current folder rather than create a separate folder config with the project. By default, if you don't add ., then Django will create an additional folder called config in which the following project structure will be created. For better understanding, you can test the command with and without . to get a clear idea of how it impacts the structure.

After executing these commands, we should be able to see the project structure shown here:

Figure 1.1: Expected project structure after executing the commands

Now that our project structure is ready, let's run `python manage.py runserver` to verify our Django project. We should see the following output in our shell:

```
python manage.py runserver
Watching for file changes with StatReloader
Performing system checks...

System check identified no issues (0 silenced).

You have 18 unapplied migration(s). Your project may not work properly until you
apply the migrations for app(s): admin, auth, contenttypes, sessions.
Run 'python manage.py migrate' to apply them.
November 04, 2023 - 18:00:26
Django version 4.2, using settings 'config.settings'
Starting development server at http://127.0.0.1:8000/
Quit the server with CONTROL-C.
```

Figure 1.2: The python manage.py runserver command's output in the shell

Please ignore the unapplied migrations warning stating **You have 18 unapplied migrations(s)** displayed in red in the console; we shall discuss this in detail in the next chapter when we learn more about the database, models, and migrations.

Now, go to your browser and open `http://localhost:8000` or `http://127.0.0.1:8000` (if the former fails to load). We shall see the following screen as shown in *Figure 1.3*, which verifies that our server is running successfully:

> **Please note**
>
> You can use `http://localhost:8000` or `http://127.0.0.1:8000` to open the Django project in your browser. If you face any error for `http://localhost:8000`, then please try using `http://127.0.0.1:8000` for all the URLs mentioned in this book.

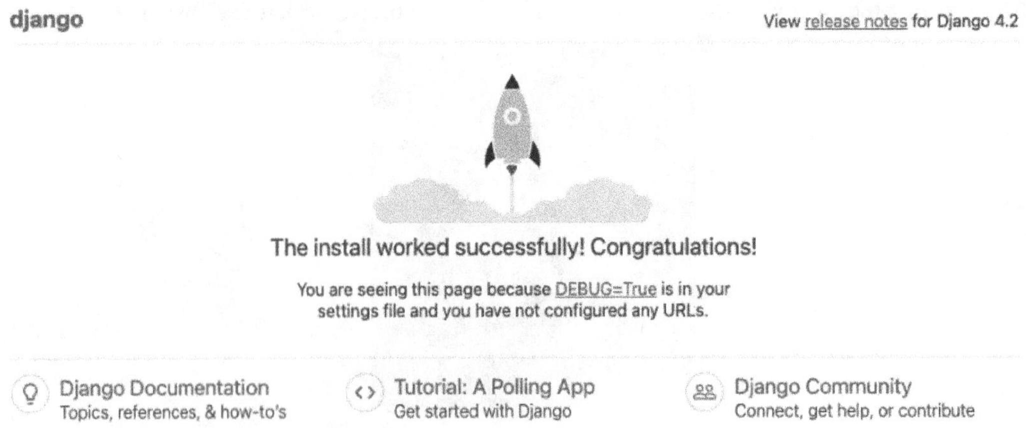

django   View release notes for Django 4.2

The install worked successfully! Congratulations!

You are seeing this page because DEBUG=True is in your
settings file and you have not configured any URLs.

Django Documentation   Tutorial: A Polling App   Django Community
Topics, references, & how-to's   Get started with Django   Connect, get help, or contribute

Figure 1.3: Our Django server running successfully with port 8000

Now, let's create our first `hello_world` view. To do this, follow these steps:

1. Open the `config/urls.py` file.

2. Add a new `view` function in `hello_world`.

3. Link the `hello_world` view function to the `hello-world` path.

   Our `config/urls.py` file should look like the following code snippet:

   ```
   from django.contrib import admin
   from django.http import HttpResponse
   from django.urls import path

   def hello_world(request):
       return HttpResponse('hello world')

   urlpatterns = [
       path('admin/', admin.site.urls),
       path('hello-world/', hello_world)
   ]
   ```

4. Open `http://127.0.0.1:8000/hello-world/` to get the result shown in *Figure 1.4*:

Figure 1.4: http://127.0.0.1:8000/hello-world/ browser response

So far, we have seen how to create the project folder structure and create our first view in Django. The example we have used is one of the smallest Django project examples that doesn't involve an app. So, let's see how we can create apps in Django that can help us manage our project better.

Creating our first app in Django

A Django app can be considered a small package performing one individual functionality in a large project. Django provides management commands to create a new app in a project; these are built-in commands that are used to perform repetitive and complex tasks. The Django community loves management commands since they take away a lot of manual effort and encapsulate a lot of complicated tasks, such as migrations and more. We shall learn more about Django management commands in the following chapters, where we will create a custom management command. However, whenever you see a command followed by `manage.py`, that is a Django management command.

So, let's create a new demo_app using the Django management command interface:

```
python manage.py startapp demo_app
```

Running this command will create the folder structure shown here:

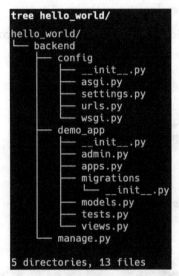

```
tree hello_world/

hello_world/
└── backend
    ├── config
    │   ├── __init__.py
    │   ├── asgi.py
    │   ├── settings.py
    │   ├── urls.py
    │   └── wsgi.py
    ├── demo_app
    │   ├── __init__.py
    │   ├── admin.py
    │   ├── apps.py
    │   ├── migrations
    │   │   └── __init__.py
    │   ├── models.py
    │   ├── tests.py
    │   └── views.py
    └── manage.py

5 directories, 13 files
```

Figure 1.5: Project structure with demo_app added

We can see that a demo_app folder has been created that contains the boilerplate code generated by Django for a new app.

> **Important note**
>
> One important step we must do whenever we create a new app is to tell Django about the new app. Unfortunately, this doesn't happen automatically when you create a new app using the Django management command. It is a manual process where you need to add the details of the new app in the INSTALLED_APPS list in the settings.py file. Django uses this to identify all the dependency apps added to the project and check for any database-related changes or even register for signals and receivers.

Though adding a new app to the INSTALLED_APPS list is not required for us currently, since we are not using *models* for Django to automatically identify any database-related changes, it is still good practice to do so. Our INSTALLED_APPS list should look like this:

```
INSTALLED_APPS = [
    ...
    'django.contrib.staticfiles',
    'demo_app',
]
```

Remember to put a comma (,) after every entry of a new app; this is one of the most common mistakes developers make and it causes Django to append two app names into one and generate a syntax error before finally correcting it.

> **Important note**
>
> In Django, third-party app integrations are also done via `INSTALLED_APPS`, so we shall see best practices around how to maintain `INSTALLED_APPS` in the following sections.

Now that we have created a new Django app with the boilerplate code, we can link the app view with `urls.py`.

Linking app views using urls.py

In this section, we shall link `views.py`, which was created by the Django management command. `views.py` is used to add business logic to the application endpoints. Just like we added the `hello_world` functional view in the previous section, we can add different functional or class-based views in the `views.py` file.

Let's create a simple `hello_world` functional view in our `demo_app/views.py` file:

```
from django.http import HttpResponse

def hello_world(request, *args, **kwargs):
    return HttpResponse('hello world')
```

As our project grows and the number of apps increases, our main `urls.py` file will become more and more cluttered, with hundreds of URL patterns in a single file. So, it is favorable to break down the main `config/urls.py` file into smaller `urls.py` files for each app, which improves the maintainability of the project.

Now, we will create a new file called `backend/demo_app/urls.py` where we shall add all the routes related to `demo_app`. Subsequently, when we add more apps to the project, we shall create individual `urls.py` files for each app.

> **Important note**
>
> The `urls.py` filename can be anything, but we are keeping this as-is to be consistent with the Django convention.

Add the following code inside the `backend/demo_app/urls.py` file:

```
from django.urls import path
from demo_app import views
```

```
urlpatterns = [
    path('hello-world/', views.hello_world)
]
```

Here, we are defining the URL pattern for the `hello-world` path, which links to the basic functional view we created earlier.

> **Opinionated note**
> We are using absolute import to import our `demo_app` views. This is a convention we shall follow throughout this book and we also recommend it for other projects. The advantage of using absolute import over relative import is that it is straightforward and clear to read. With just a glance, someone can easily tell what resource has been imported. Also, PEP-8 explicitly recommends using absolute imports.

Now, let's connect the `demo_app/urls.py` file to the main project `config/urls.py` file:

```
from django.contrib import admin
from django.urls import include
from django.urls import path

urlpatterns = [
    path('admin/', admin.site.urls),
    path('demo-app/', include('demo_app.urls'))
]
```

Next, open `http://127.0.0.1:8000/demo-app/hello-world/` in your browser to make sure our `demo-app` view is linked with Django. You should be able to see `hello world` displayed on the screen, just as we saw earlier in *Figure 1.4*.

So far, we have worked with plain vanilla Django, but now, we'll see how we can integrate DRF into our project.

Integrating DRF

In the API-first world of development, where developers create APIs day in, day out for every feature they build, DRF is a powerful and flexible toolkit for building APIs using Django.

> **Important note**
> If you are not familiar with the basics of DRF, we will be going through the basics in this book. However, you can find more information here: `https://www.django-rest-framework. org/tutorial/quickstart/`.

Now, let's integrate DRF into our `hello_world` project. First, we need to install DRF in the virtual environment:

```
pip install djangorestframework
```

Now, go to `settings.py` and add `rest_framework` to `INSTALLED_APPS`. As you may recall, when we were integrating `demo_app` into the project, we mentioned that as the project grows, the `INSTALLED_APPS` list will also grow. To manage this better, we shall split our `INSTALLED_APPS` list into three sections:

- `DJANGO_APPS`: This will contain a list of all the default Django apps and any new Django built-in apps we shall add to the project

- `THIRD_PARTY_APPS`: Here, we shall maintain all the third-party apps we are integrating into the project, such as `rest_framework`

- `CUSTOM_APPS`: We shall add all the apps we are creating for the project to this list – in our case, `demo_app`

Here is an example of how your `INSTALLED_APPS` list will look in the `settings.py` file:

```
DJANGO_APPS = [
    'django.contrib.admin',
    'django.contrib.auth',
    'django.contrib.contenttypes',
    'django.contrib.sessions',
    'django.contrib.messages',
    'django.contrib.staticfiles',
]
THIRD_PARTY_APPS = [
    'rest_framework',
]

CUSTOM_APPS = [
    'demo_app',
]

INSTALLED_APPS = DJANGO_APPS + CUSTOM_APPS + THIRD_PARTY_APPS
```

So far, we have been using Django `HttpResponse`. Now, we shall integrate the DRF response into our view. So, go to the `demo_app/views.py` file and add the following code:

```
from rest_framework.decorators import api_view
from rest_framework.response import Response

@api_view(['GET'])
```

```
def hello_world_drf(request, *args, **kwargs):
    return Response(data={'msg':'hello world'})
```

The integration of the DRF function-based view will change the UI completely for our endpoint. If you open http://127.0.0.1:8000/demo-app/hello-world-drf/, it will have a much more verbose UI, giving us a lot more information than before. This is particularly helpful when we start working with HTTP requests other than GET requests.

Here is our basic Django project integrated with DRF:

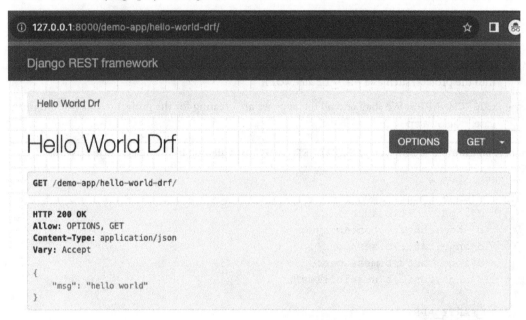

Figure 1.6: Using the DRF response in our hello-world-drf view

Now that we have a working project in Django, let's learn some good practices that can help you go the extra mile without making mistakes yourself, rather than learning from the ones you have made earlier. DRF is the most popular Django package as it helps developers create REST endpoints. Now, let's see what some good practices are for creating REST APIs.

Creating RESTful API endpoints with DRF

The most popular and widely used API is the REST API. Throughout this book, we shall be working with the REST API. REST has been around for more than two decades, and every company has its interpretation and implementation. In the following section, we shall try to put all the best practices used in the industry into practice.

> **Opinionated note**
>
> The RESTful API is not a protocol; instead, it is a standard convention that developers follow. There is no right or wrong while designing RESTful APIs. Since there is no enforced standard, the details we will provide are purely opinionated and come from my past experiences. You are free to pick the points you like and leave out the things that you feel are not relevant to your implementations.

Best practices for defining RESTful APIs

Let's look at a few generic good practices that developers use in the industry while defining RESTful endpoints:

- Using nouns instead of verbs in endpoint paths using appropriate HTTP request methods. Here are some examples (*please note that the URL example used here is just an outline of how we should define our REST URLs and that we are not defining the exact code*):

```
# To get all blogs
Avoid GET /get-all-blogs, rather use GET /blogs
# To delete a particular blog
Avoid POST /delete-blog rather use DELETE /blogs/<blogId>
# To create a new blog with POST request
Avoid POST /create-new-blog rather use POST /blogs
# To update an existing blog with a PUT request
Avoid PUT /update-blog rather use PUT /blogs/<blogId>
```

- Using the appropriate HTTP method is preferred to perform CRUD operations. There are multiple HTTP methods present, but we shall only cover the top five commonly used methods:

    - GET: To retrieve an entity, be it a list or detail

    - POST: To create any new entity

    - PUT: To Update an entity

    - PATCH: To partially update an entity

    - DELETE: To delete an entity

- It is preferred to create plural nouns in the endpoint. When you have to get a single entry, then use id after the endpoint to retrieve the information. For example, to get a list of blogs, use GET /blogs, and to get the details of one blog, use GET /blogs/<blog id>.

- Using a logical nested structure for an endpoint is important to clean the API interface and maintain a good information architecture. For example, to get all the comments for a particular blog, the API should be GET /blogs/<blog id>/comments.

- Versioning the API is important since it helps support legacy systems without breaking the contract in newer systems. Examples of this are `/v1/blogs/` and `/v2/blogs`. We will learn more about this later, in the *Using API versioning* section.

- Servers should send appropriate HTTP response status codes as per the action, along with the message body (if applicable). Here are a few of the most widely used HTTP status codes:

  - **2xx**: Used for any success. For example, `200` is for any request responding with the data successfully, `201` is for creating a new entry, and so on.

  - **3xx**: Used for any redirection.

  - **4xx**: For any error. For example, use `400` for bad requests and `404` for requested data not found.

  - **5xx**: When the server crashes due to any unexpected request or the server is unavailable.

- The server must accept and respond with a JSON response. The API will not support other data types, such as plain text, XML, and others.

Best practices to create a REST API with DRF

DRF is a framework that helps us create the REST endpoint faster. It's the responsibility of the developer to write scalable and maintainable code while following the best practices. Let's look at some best practices that we can implement using DRF.

Using API versioning

Creating versions of an API is probably the most important thing to follow when working with clients whose updates are not under your control. An example of this is working on a mobile app. Once an end user installs a given mobile app, with a given API integrated, we have to support the given API until the end user updates the mobile app version with the newer API.

While creating an endpoint, a developer should consider all the future requirements possible, along with all the corner cases. However, just like it is not possible to predict the future, a developer cannot always foresee how the current API design might have to be redesigned. A redesign would mean breaking the contract between the client and the server. This is when the importance of API versioning comes into the picture. A well-versioned API will implement a new contract without breaking any of the existing clients.

There are multiple ways to implement API versioning:

- **Accept header versioning**: Since the version is passed through the `Accept` header, whenever there is a new version, the client doesn't need to update any endpoint whenever a new version is created:

```
GET /bookings/ HTTP/1.1
Host: example.com
Accept: application/json; version=1.0
```

- **URL path versioning**: The API version is passed through the URL path pattern. This is one of the most widely used API versioning methods due to it providing better visibility:

```
GET /v1/bookings/ HTTP/1.1
Host: example.com
Accept: application/json
```

- **Query parameter versioning**: The query parameter in the URL contains the version. After URL path-based versioning, this is the second most common versioning method due to its cleaner interface and better discoverability:

```
GET /something/?version=1.0 HTTP/1.1
Host: example.com
Accept: application/json
```

- **Host name versioning**: This involves using the subdomain to pass the API version in the hostname. This kind of versioning is used when someone migrates the whole service to a newer version rather than single endpoints:

```
GET /bookings/ HTTP/1.1
Host: v1.example.com
Accept: application/json
```

DRF supports all four methods of API versioning out of the box and also gives the option to create our custom API version logic if needed. We shall explore **URLPathVersioning** primarily since it is one of the easiest and most popular ways of implementing versioning using DRF.

In URL path versioning, the API version is passed in the URL path, which makes it easy to identify on both the client and server side. To integrate URLPathVersioning in DRF, add the following in the Django config/settings.py file:

```
REST_FRAMEWORK = {
    'DEFAULT_VERSIONING_CLASS': 'rest_framework.versioning.
URLPathVersioning'
}
```

Now, we must add the version to the URL path. It is important to name the URL parameter <version> since DRF is expecting it to be <version> by default. Here, <version> is the URL's pattern, which means that any URL that matches this pattern shall be linked to the views.

> **Important note**
> To learn more about urlpatterns, go to https://docs.djangoproject.com/en/stable/topics/http/urls/.

It is advisable to add <version> at the beginning of the URL, so let's do that in the main config/urls.py file:

```
urlpatterns = [
    path('admin/', admin.site.urls),
    path('<version>/demo-app-version/', include('demo_app.urls'))
]
```

Once we have configured the URL with the <version>, we can try to create a new view and retrieve the version in our view. Add the following code to your demo_app/urls.py file:

```
from django.urls import path

from demo_app import views

urlpatterns = [
    path('hello-world/', views.hello_world),
    path('demo-version/', views.demo_version),
]
```

We shall retrieve the API version in the view and return the version in response:

```
@api_view(['GET'])
def demo_version(request, *args, **kwargs):
    version = request.version
    return Response(data={
        'msg': f'You have hit {version} of demo-api'
    })
```

Now, when we open http://127.0.0.1:8000/v1/demo-app-version/demo-version/, we should be able to see the following screen:

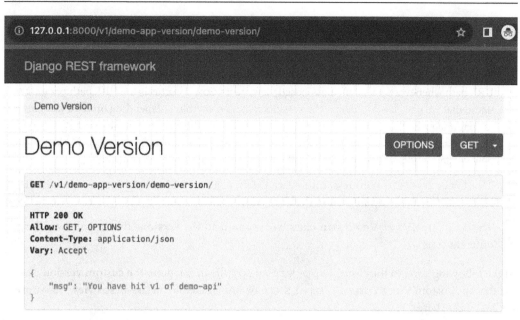

Figure 1.7: Output showing which version we have hit for the given API

If we change the URL to http://127.0.0.1:8000/v9/demo-app/demo-version/, then we'll see that it returns v9. v9 might not have been released yet, so this might create confusion for the end user hitting the endpoint. To solve this problem, we shall see how we can customize the version class of DRF so that we can add constraints that can help us design better applications.

Using a custom version class with DRF

Let's see how we can extend the URLPathVersioning class provided by DRF to address the problem we just raised. First, create a file called demo_app/custom_versions.py. This file will have a custom version class for each view, along with a default class for all the views that don't have multiple versions yet:

```python
from rest_framework.versioning import URLPathVersioning

class DefaultDemoAppVersion(URLPathVersioning):
    allowed_versions = ['v1']
    version_param = 'version'

class DemoViewVersion(DefaultDemoAppVersion):
    allowed_versions = ['v1', 'v2', 'v3']

class AnotherViewVersion(DefaultDemoAppVersion):
    allowed_versions = ['v1', 'v2']
```

Let's see what the preceding code does:

- The `DefaultDemoAppVersion` class can be used for all the views that are created in `demo_app`. It has an `allowed_versions` attribute that lists all the allowed versions that can be used in the URL path whenever we use this class-based view. `version_param` is the URL path parameter name that we have used to define the version; it can be anything, depending on how you name the parameter, but in our case, we are using `<version>`, which is used in the `config/urls.py` file. This class will be used for all the views that are created in the demo app by default until a new version is added, after which we will create an individual class, as shown next.

- The `DemoViewVersion` class will contain the list of all the `allowed_versions` attributes for `DemoView` that are allowed in the URL path.

- The `AnotherViewVersion` class will contain all the versions that are allowed for a different class.

Add the following code to the `demo_app/views.py` file to integrate the custom version class (note that the custom `versioning_class` can be only linked to a class-based view, so we are using `APIView` here):

```python
from rest_framework.response import Response
from rest_framework.views import APIView

from demo_app import custom_versions

class DemoView(APIView):
    versioning_class = custom_versions.DemoViewVersion

    def get(self, request, *args, **kwargs):
        version = request.version
        return Response(data={'msg': f' You have hit {version}'})

class AnotherView(APIView):
    versioning_class = custom_versions.AnotherViewVersion

    def get(self, request, *args, **kwargs):
        version = request.version
        if version == 'v1':
            # perform v1 related tasks
            return Response(data={'msg': 'v1 logic'})
        elif version == 'v2':
            # perform v2 related tasks
            return Response(data={'msg': 'v2 logic'})
```

Let's explore the code and understand what is happening under the hood when we use the custom version class:

- The DemoView class is a class-based APIView where we are passing the allowed versions for the view by the versioning_class attribute. This allows the request object to have a version attribute that is parsed from the URL path. Since we have specified the DemoViewVersion class, this view will only allow the v1, v2, and v3 versions in the URL path. Any other version in the path will result in a 404 response.

- The AnotherView class is a class-based view where we are passing AnotherViewVersion as the versioning_class attribute. In this view, we are bifurcating the request by checking different versions and responding differently whenever we have a v1 or v2 request.

Now, to link the view logic to the demo_app/urls.py file, add the following code:

```
urlpatterns = [
    path('hello-world/', views.hello_world),
    path('demo-version/', views.demo_version),
    path('custom-version/', views.DemoView.as_view()),
    path('another-custom-version/', views.AnotherView.as_view())
]
```

If we go to http://127.0.0.1:8000/v4/demo-app-version/custom-version/ in our browser, we shall see the following error as shown in *Figure 1.8*, since we have only allowed three versions in our custom versioning_class:

Figure 1.8: 404 error message stating "Invalid version in URL path"

This serves our purpose of only allowing certain versions of the API; any other API version shall result in an error response.

> **Important note**
>
> Custom versioning can only be attached to class-based views. If you don't pass any custom `versioning_class`, then Django will pick `DEFAULT_VERSIONING_CLASS` from the default settings.

Avoid Router

Frameworks such as Ruby on Rails provide the functionality to automatically map requests to a given pattern of URLs, depending on the functionality. DRF borrowed this concept and incorporated it into the framework as the Routers feature. Though this is a wonderful concept to learn and experiment with, developers should avoid it in production since this goes against the principle of Django: "*Explicit is better than implicit.*"

Django mentions that it should not show too much of the magic. I have personally seen legacy systems where developers added Router and after a couple of months, when a different developer wanted to fix a bug in the view, they were unable to find the corresponding view directly before having the "aha!" moment of identifying the Router concept.

> **Opinionated note**
>
> Avoiding the use of Router is something I have learned the hard way and have seen multiple developers avoid in production. But this is also an opinion that was developed through a bad experience; you can always try to implement it in a better way in your project.
>
> If you want to learn more about Router, you can do so here: `https://www.django-rest-framework.org/api-guide/routers/`.

With that, we've learned how to create RESTful APIs and work with versioning. Now, let's learn how to work with views using DRF. We mainly write business logic inside views.

Working with views using DRF

DRF extends the views concept of Django to provide a better interface for creating REST endpoints. Just like functional and class-based views in Django, DRF also supports both of them. However, it depends on the developer to choose which type of view fits their use case.

As a rule of thumb, I am always inclined toward functional views when I have a standalone endpoint where the logic is straightforward and won't have any complexity in the future. Class-based views have a learning curve that stops developers from using them initially, but once the entry barrier is breached, developers rarely move back to functional views. Let's explore both types of views in more detail.

Functional views

Django provides an easy functional view interface that helps any developer get going with faster development; that is why it is very popular for any developer starting Django. DRF keeps the simplicity of Django views and allows developers to convert a Django view into DRF using a simple decorator, `@api_view`. I prefer using it when the logic is straightforward or when I'm working on a simple project that will not get too complex.

DRF gives us `@api_view` to work with regular Django functional views. Adding the decorator converts the usual Django `HttpRequest` into a `Request` instance. The decorator takes a list of all the HTTP-allowed methods for the given functional view block; any method that is not listed will not be allowed. By default, if no HTTP method is mentioned, then it will only allow `GET` methods. For a response to views, we should use the `Response` class from DRF rather than `HttpResponse` from Django; it automatically takes care of bypassing **cross-site request forgery** (**CSRF**) for views and gives us a UI to interact with the backend service.

Let's look at an example of a DRF functional view:

```
from rest_framework.decorators import api_view
from rest_framework.response import Response


@api_view(['GET', 'POST', 'PUT'])
def hello_world(request, *args, **kwargs):
    if request.method == 'POST':
        return Response(data={'msg': 'POST response block'})
    elif request.method == 'PUT':
        return Response(data={'msg': 'PUT response block'})
    return Response(data={'msg': 'GET response block'})
```

Now, let's learn how to work with class-based views, which are widely used throughout the industry.

Class-based views

As mentioned previously, class-based views have a learning curve, so they're avoided initially by a lot of developers. Even I avoided it for a couple of months until I saw the bigger picture, but ever since, there has been no looking back. It takes advantage of the inheritance property and helps implement the **Don't Repeat Yourself** (**DRY**) principle. For large code bases, it is an absolute must, and I recommend that anyone starting a new project implement it from the start.

DRF provides two types of class-based views: `APIView` and Generic Views.

APIView

The DRF **APIView** class is an extension of Django's View class. Using the APIView class converts the default Django HttpRequest into a Request object, and the handler methods can return DRF's Response object rather than Django's HttpResponse. It supports additional policy attributes such as authentication_classes, permission_classes, and versioning_classes, which make the life of a developer much easier. We shall use them in the following chapters and discuss them in detail.

The interface for using APIView is simple: to implement any HTTP request method for a given endpoint, simply create a new method in the class with the given name; for example, a GET request would have a get method, and a DELETE request would have a delete method, and so on. Whenever a request is hit for the given HTTP method, the corresponding method will be called automatically.

Let's implement a quick example with APIView to learn more:

```python
from rest_framework.views import APIView


class DemoAPIView(APIView):
    def get(self, request, *args, **kwargs):
        return Response(data={'msg': 'get request block'})

    def post(self, request, *args, **kwargs):
        return Response(data={'msg': 'post request block'})

    def delete(self, request, *args, **kwargs):
        return Response(data={'msg': 'delete request block'})
```

Linking an APIView implemented class to the urls.py file is different from how we linked functional views earlier. Instead, we use <class name>.as_view() to link it to the corresponding URL:

```python
urlpatterns = [
    ...
    path('apiview-class/', views.DemoAPIView.as_view())
]
```

We shall expand our knowledge of APIView more as we learn about different concepts surrounding DRF and Django.

Generic Views

While building a web application, there comes a point when developers are doing the same monotonous job of writing repetitive logic. That is when the principle of DRY kicks in and the developer thinks about how to solve this repetitive pattern. The answer to this is **Generic Views**.

Generic Views is one of the most widely popular features of Django and DRF that helps developers build a basic CRUD operation API at lightning speed. Generic Views are tightly coupled with DRF's serializer and model concepts, so we shall discuss Generic Views in detail once we learn about these concepts in the following chapters.

Opinionated note

We are not discussing **Viewsets** here since it is primarily used with **Router**. We have already discussed why we should avoid Router in Django, but if you are keen to learn more about it, you can go to the DRF documentation, which explains it quite well with relevant examples.

For more details about `Viewsets`, see `https://www.django-rest-framework.org/api-guide/viewsets/`.

Now that we have seen the basic integration of DRF and how to work with views, let's focus on how to improve the development experience. Web development is more than just writing code – it is also about using the right tools. Throughout this book, we shall introduce more and more tools for different purposes. For now, let's learn about API development tools.

Introducing API development tools

While developing APIs, it is important to make sure the endpoint is working with every possible input and that we manually test the APIs before handing them over to the client team for integration. Though it is always favorable to write unit and integration tests for every piece of code/API you create, we might not be able to write them. Due to this, API development tools come in handy to support you through this journey.

Postman is one of the most popular tools in the developer community that helps developers build APIs at every stage of the development process. Though it is common to use Postman for testing APIs, it has far more use cases than just testing the APIs, as follows:

- Creating mock servers with the API contract to help the client team work independently while the backend APIs are being developed

- Creating documentation and OpenAPI specs for the team to work with endpoints

- Creating a collection of APIs that can group APIs as per the use case

- Creating tests that can run periodically on remote servers to make sure the contract of the APIs has not changed

The tutorial series on Postman's website is quite intuitive: `https://learning.postman.com/docs/getting-started/introduction/`. Please go through the tutorial if you are not familiar with it as we shall use Postman throughout this book to test our endpoints. You can use other tools for the same purpose, such as *hoppscotch* (`https://hoppscotch.io/`) and *testmace* (`https://testmace.com/`).

Learning new tools is a skill set that is important for the growth of any developer since these tools ease our development journey. Throughout this book, we shall introduce and learn about new tools that will help with just that.

Summary

In this chapter, we learned how to set up our local development environment with Python's virtual environment and created a Django project from scratch. When we start a new project, the project structure is very important and, in this chapter, we have learned how to follow the best practices to create a new project structure for a Django web application.

Throughout this book, we will be working with REST APIs, so we provided a basic introduction to REST APIs and the best practices around how to create them, such as naming conventions, adding versioning, and using appropriate HTTP methods and status codes. We also learned how to integrate DRF with Django and create a basic demo app using URLs, functional views, and class-based views.

By reading this chapter, we should be comfortable with creating a Django web application from scratch with multiple Django apps, as well as URLs and views. We should be able to add Django apps to `settings.py` and create RESTful APIs using DRF.

So far, we have only worked with static responses without involving any databases. In the following chapter, we shall learn how to set up our database and connect a Postgres database to our Django project.

Exploring Django ORM, Models, and Migrations

2

In *Chapter 1*, we saw how to create a Django project and integrate Views into the project. All the responses in the Views were static and we didn't see how to implement any code to get dynamic responses using databases.

In this chapter, we shall see how to integrate a PostgreSQL database into our Django project and the support Django provides out of the box to work with models and databases. If you have worked with **relational database management systems** (**RDBMSs**) before, you will know how painful it is to perform migrations whenever the schema is updated. However, Django provides migration support out of the box, which abstracts all the intricacies under two management commands.

After that, we'll learn a few best practices while working with Django models and how to check for performance optimizations with large database tables. Finally, we'll check out the newest feature that was introduced in Django 4.1: Async ORM.

We'll cover the following topics in this chapter:

- Setting up PostgreSQL with a Django project
- Using models and Django ORM
- Understanding the crux of Django migrations
- Exploring best practices for working with models
- Learning about performance optimization
- Exploring Django Async ORM

Technical requirements

In this chapter, we'll work with the Django model and database. To work with the code in this chapter, you will need to have Django installed on your system with the project structure discussed in the previous chapter. You should be comfortable in creating Django apps and connecting Views to the URLs, as we have learned previously.

Apart from these, you should have the following in your skillset:

- You should understand the basics of RDBMSs and how databases work at a high level. You should also be familiar with the concepts of normalization and denormalization of databases.

- You should understand the basics of Django ORM and have worked with Django models before (you may have followed and understood the Django Girls tutorial mentioned in the previous chapter).

- You should understand the concepts surrounding SQL and have PostgreSQL installed on your system or have a stable internet connection to use a remote PostgreSQL server (`https://www.elephantsql.com/`).

- You should have TablePlus (`https://tableplus.com/`) or any database management GUI tool installed.

This book's GitHub repository, which contains all the code for this chapter, can be found at `https://github.com/PacktPublishing/Django-in-Production/tree/main/Chapter02`.

> **Important note**
>
> Database and Django ORM have a wide range of concepts, and it is not possible to cover each topic in this chapter. What I have seen from my personal experience is that many developers are unaware of trivial concepts, which prevents them from using advanced concepts in their projects and results in them making rookie mistakes. Though we shall try to explain all of the important concepts, you can look at the information boxes provided throughout this chapter, which will direct you to the official document for extra information.

> **Important note**
>
> In this chapter, we'll use snippets of Django ORM and Django ORM queries. The code snippets that are used in this chapter might not be self-sufficient and might need additional code to execute. If you are facing any problems understanding the code snippets or are unable to execute them independently, then join our *Django in Production* Discord channel and ask us there. Here is the invite link for our Discord server, where you can reach me directly: `https://discord.gg/FCrGUfmDyP`.

Setting up PostgreSQL with a Django project

Django integrates **SQLite3** for initial development purposes out of the box, which helps developers get started with their projects without the hassle of initial database setup. In our case, we used SQLite3 in *Chapter 1*. However, it is a file-based database that is not recommended to be used in production since it cannot scale.

Instead, we shall use **PostgreSQL** as our choice of database to work with Django throughout this book. Since it is always recommended to keep your local development environment as close to the production environment as possible, we shall set up PostgreSQL for our local development.

> **Important note**
>
> You can use other supported databases such as MySQL, MariaDB, and Oracle (`https://docs.djangoproject.com/en/stable/ref/databases/#databases`) by just making a small change in the settings file. But please remember that not all databases support all the ORM functionalities, so if you are using another database, check the documentation for more details.

Setting up PostgreSQL in a local system can sometimes be a daunting task, so I recommend using Docker for this purpose (we'll learn more about it in *Chapter 11*). However, for ease of learning, we shall use ElephantSQL, a remote PostgreSQL server, to get started. ElephantSQL (`https://www.elephantsql.com/`) provides a free PostgreSQL server with limited configuration; this is sufficient for you to follow everything we'll learn about in this book.

So, let's see how we can create a PostgreSQL server using ElephantSQL, which will then be used to configure Django to use PostgreSQL.

Creating a PostgreSQL server

If you already have a PostgreSQL server, you can skip to the *Configuring Django with PostgreSQL* section. If not, follow these steps to create your free PostgreSQL server using ElephantSQL:

1. Go to `https://customer.elephantsql.com/login` and create a new free account (a credit card isn't required, so anyone can create one).

2. Once you've logged in, you will see the following screen. Click **Create New Instance**:

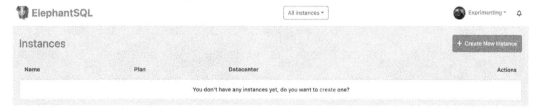

Figure 2.1: ElephantSQL – new user dashboard

3. Next, fill in the following details:

    A. **Name**: The name of the instance; you can choose any name that you feel is suitable

    B. **Plan**: Select **Tiny Turtle (Free)**

    C. **Region**: You can select any region and cloud provider

4. After this, you should be able to see your database instance on the ElephantSQL dashboard. *Figure 2.2* shows my database configurations:

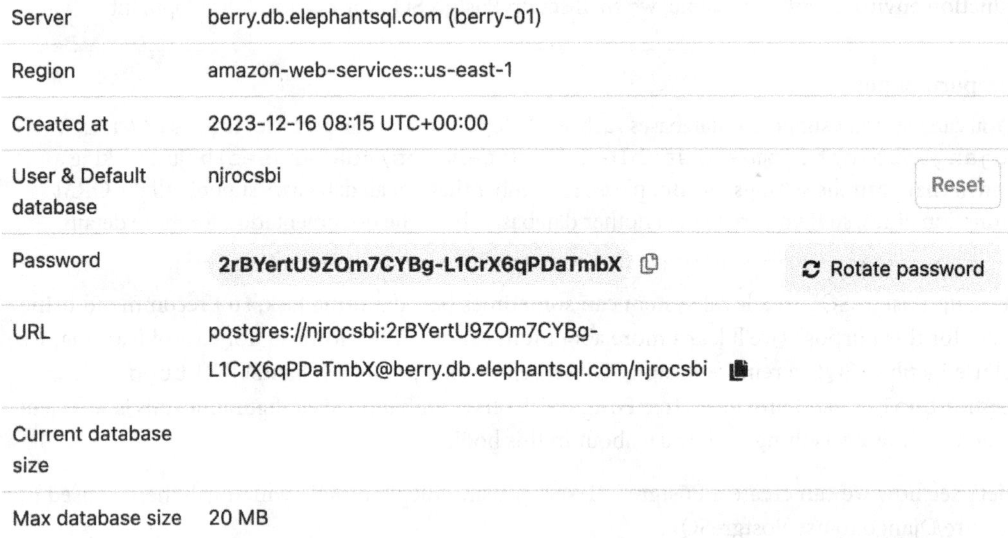

Server	berry.db.elephantsql.com (berry-01)	
Region	amazon-web-services::us-east-1	
Created at	2023-12-16 08:15 UTC+00:00	
User & Default database	njrocsbi	Reset
Password	**2rBYertU9ZOm7CYBg-L1CrX6qPDaTmbX** 📋	↻ Rotate password
URL	postgres://njrocsbi:2rBYertU9ZOm7CYBg-L1CrX6qPDaTmbX@berry.db.elephantsql.com/njrocsbi 📋	
Current database size		
Max database size	20 MB	

Figure 2.2: PostgreSQL server details, as shown in the ElephantSQL dashboard

Take note of the **Server**, **User & Default database**, and **Password** fields as we'll use these in the next step.

5. Now, we'll try to connect to the PostgreSQL server using TablePlus (`https://tableplus.com/`) or any other database GUI tool with port `5432`. Use the information you've just gathered to fill in the details in the connection wizard and hit **Test**; you should see the following screen:

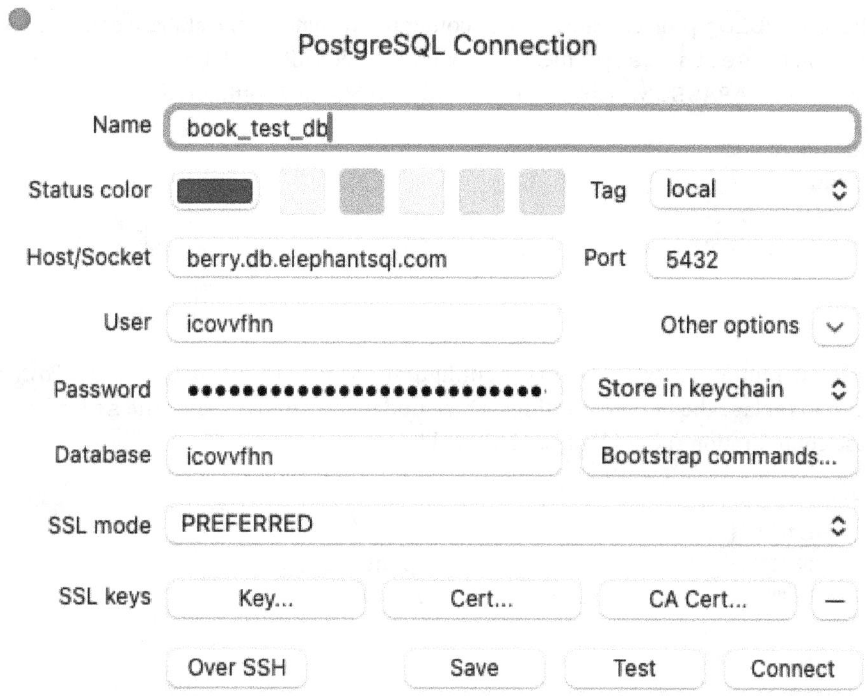

Figure 2.3: The PostgreSQL Connection wizard screen of TablePlus when the test connection is successful

> **Important note**
>
> By default, PostgreSQL runs on port 5432. ElephantSQL also exposes the same port for end users to connect through different applications.

6. Once the fields are all green, you can hit the **Connect** button to open the database in your local GUI.

Now that we have established a connection to the remote PostgreSQL server from our local system, we'll learn how to configure Django so that it can connect to PostgreSQL.

Configuring Django with PostgreSQL

In the previous chapter, we learned how to create the Django project structure; we'll follow the same structure going forward. Throughout this book, we'll use an example of a blogging app to understand different in-depth features of Django. You can check out the project structure at https://github.com/PacktPublishing/Django-in-Production.

Once you have a `myblog` project ready, we can configure the database. Database configurations in Django are put in the `settings.py` file. If you open your default `settings.py` file, you will be able to locate the `DATABASES` variable, which already has SQLite3 configured:

```
DATABASES = {
    'default': {
        'ENGINE': 'django.db.backends.sqlite3',
        'NAME': BASE_DIR / 'db.sqlite3',
    }
}
```

Now, let's use the remote PostgreSQL server configurations we saw in the *Creating a PostgreSQL server* section to replace the SQLite3 settings. Once updated with the values, the `settings.py` file's database configuration should look something like this:

```
DATABASES = {
    'default': {
        'ENGINE': 'django.db.backends.postgresql',
        'HOST': 'tiny.db.elephantsql.com',
        'NAME': 'wcgzvqra',
        'USER': 'wcgzvqra',
        'PORT': 5432,
        'PASSWORD': 'lyuX6vWvAtrqZSY1oiPKNBMoKFkctBVn',
    }
}
```

After we add the PostgreSQL configuration, run `python manage.py runserver` in the terminal. This will give us a long error stating `No module named 'psycopg2' is found`. To fix this, Django needs an additional package called `'psycopg2'`, which is a PostgreSQL adapter for Python.

We are not using `psycopg2` in this book as it needs to be built using the source code. Instead, we will be using `binary` since it is recommended for beginners. Let's install this package using the following command:

```
pip install psycopg2-binary
```

This will install the PostgreSQL client package in our local Python environment.

> **Important note**
> Use `psycopg2-binary` during development, but try to avoid it during production. For more information, see `https://www.psycopg.org/docs/install.html#psycopg-vs-psycopg-binary`.

Now, run the `python manage.py runserver` command again; we should see the default lift-off screen that we get when we open `http://127.0.0.1:8000` on our browse (just like we saw in *Chapter 1*).

> **Important note**
>
> Hardcoding a database or any project-related configuration in the `settings.py` file isn't the recommended or optimal way. We should use environment variables to perform such operations. We'll learn more about using environment variables to fetch project configuration while working with Docker in *Chapter 11*.

Now that we have PostgreSQL integrated into our Django project, let's learn how to work with models and Django ORM.

Using models and Django ORM

For any organization, data is one of the most important foundational elements in Django, and the `models.py` file is used to structure how data is stored in the database. A well-architected system would have a strong Django model foundation, where the developer has spent enough time and done proper research before creating the database schema.

One of the most common questions developers have while designing the schema is whether to perform normalization or denormalization. The answer is always "it depends." It depends on multiple factors, such as the data access pattern and the performance bottleneck we are trying to solve. So, as a rule of thumb, it is always preferred to start with a normalized database schema design and then, as we find different use cases, try to solve the problem by introducing caching or indexes, and then finally perform denormalization.

> **Important note**
>
> We won't focus on the basics of how to work with Django models or how to write queries using Django ORM. If you are not familiar with the basics of Django models, please go through the Django tutorial on the official website or read the official documentation at `https://docs.djangoproject.com/en/stable/ref/models/querysets/`.

We won't be covering the basic concepts of Django ORM, but in the next section, we'll integrate our Django models into the blogging application.

Adding Django models

For different blogging applications, we can have different types of database models as per the features. In our application, we'll have basic features such as `blog` and `author` to begin with. We will be creating separate Django apps to add these models:

```
python manage.py startapp author
python manage.py startapp blog
```

Now, add apps to `settings.py` in the CUSTOM_APPS section:

```
CUSTOM_APPS = [
    'blog',
    'author',
]
```

Now, to go the `author/models.py` file and add the `Author` model:

```
class Author(models.Model):
    name = models.CharField(max_length=100)
    email = models.EmailField()
    bio = models.TextField()

    def __str__(self):
        return self.name
```

Similarly, add the `Blog` model to the `blog/models.py` file:

```
class Blog(models.Model):
    title = models.CharField(max_length=100)
    content = models.TextField()
    author = models.ForeignKey('author.Author', on_delete=models.
CASCADE)
    created_at = models.DateTimeField(auto_now_add=True)
    updated_at = models.DateTimeField(auto_now=True)

    def __str__(self):
        return self.title
```

Now, to run the migration command to create the migrations, apply the migrations to the database:

```
python manage.py makemigrations
python manage.py migrate
```

With that, we have successfully added a basic database model using Django ORM. Now, let's discuss a few important points about different Django ORM functionalities so that we have a better insight into working efficiently with Django.

Basic ORM concepts

In this section, we'll learn about different ORM concepts that help Django developers implement business logic using ORM efficiently.

Null versus blank

When creating a new model field, `null=True` signifies that the given column is allowed to store a `null` value, while `blank=True` means that Django Admin will allow the field to have an empty string as a valid value. In short, `null=True` is a database constraint, while `blank=True` is a Django application constraint.

We can also use `null` and `blank` together. For example, we can allow the `bio` field for the `Author` model to be `null` as well as `blank`:

```
class Author(models.Model):
    ...
    bio = models.TextField(null=True, blank=True)
```

auto_now versus auto_now_add

The `DateTime` field has two options that allow developers to automatically save the current time to the database field:

- `auto_now` updates the value with the current `DateTime` field whenever someone makes an edit to a given row
- `auto_now_add` only adds the current timestamp when the object is created

Now, an important question is, when should we use them? As a basic rule of thumb, I recommend using `auto_now_add` to capture a *created at* timestamp and `auto_now` to capture the last *updated at* timestamp.

Let's learn how to use these options in Django ORM:

```
class Blog(models.Model):
    ...
    created_at = models.DateTimeField(auto_now_add=True)
    updated_at = models.DateTimeField(auto_now=True)
    ..
```

Avoid using raw SQL queries

Django ORM is a wrapper on top of database queries, so, at times, developers will be tempted to add SQL query logic directly. Django gives us different interfaces, such as the `connection.cursor` context manager and the `<Model>.objects.raw()` model ORM interface, to execute raw queries through a Django database connection. Try to avoid using raw SQL queries in Django and, unless necessary, make sure that you have enough checks in your application code to prevent SQL injection.

Use query expressions and database functions

Query expressions are one of the most under-utilized features that developers avoid due to their learning curve. While building application logic, you should always try to perform all the mathematical and logical operations in SQL. For example, if an application use case needs us to find the average salary for employees, we should avoid writing the logic in Python; instead, we should use Django ORM's `aggregate` to perform this operation – for example, `<model>.objects.aggregate(Avg('salary'))`.

> **Important note**
>
> The Django documentation provides a wide range of examples of query expressions and database functions, such as *aggregate*, *F*, and *window* functions, which help with performing computation in the database layer. This is faster and more scalable than regular logic written in Python.
>
> Please go through the following documentation to learn more about query expressions and database functions so that you can use the Django ORM layer better. For query expressions, check out `https://docs.djangoproject.com/en/stable/ref/models/expressions/`. For database functions, check out `https://docs.djangoproject.com/en/stable/ref/models/database-functions/`.

Use a reverse foreign key lookup

When we create a foreign key relationship between Model A and Model B, a reverse relationship is also established. For example, let's consider the `Author` and `Blog` model:

```
class Author(models.Model):
    name = models.CharField(max_length=100)
    email = models.EmailField()
    bio = models.TextField()


class Blog(models.Model):
    title = models.CharField(max_length=200)
    content = models.TextField()
    author = models.ForeignKey('author.Author', on_delete=models.
PROTECT)
```

In the preceding code, we defined the `Author` model and the `Blog` model, which are linked by the `ForeignKey` field. Now, let's learn how to use foreign keys in our application.

> **Setting up dummy data**
>
> Before we run the snippets in this section, we need to fill our database with dummy data. You can create dummy data yourself or follow the instructions mentioned in this book's GitHub repository: `https://github.com/PacktPublishing/Django-in-Production/tree/main/Chapter02`.

In this example, it is easy to fetch the author's information from a `blog` object by directly accessing the `author` object:

```
>>> from blog import models as blog_models
>>> author_obj = blog_models.Blog.objects.get(id=1).author
>>> author_obj

<Author: John Doe>
```

But if we have to fetch the blog information from an `author` object, we can use a reverse foreign key lookup:

```
>>> from author import models as author_models
>>> author = author_models.Author.objects.get(email='john@gmail.com')
>>> all_blogs_by_an_author = author.blog_set.all()
>>> selected_blog = author.blog_set.filter(title='Python is cool')
>>> selected_blog

<QuerySet [<Blog: Python is cool>]>
```

Here, `blog_set` is automatically named since we have not set any `related_name` attribute. We can get all the blogs by the selected author by using `.all()` and also use it as a regular QuerySet, as we have done in the `filter` option.

With that, we have learned a few basic concepts that can help us operate better with ORMs and Django models. Now, let's learn how to retrieve raw SQL queries from Django ORM.

How to get raw queries from ORM

Django provides a lot of abstraction layers on top of SQL queries, which enables us to perform any database-related operation without getting into the intricacies of SQL queries. However, at times, it is necessary to check out the raw query that Django creates under the hood to perform some optimizations or take a deeper look into the query.

While working with QuerySets, retrieving database queries is quite easy. For example, if we print the `.query` attribute of a QuerySet, then we can get the raw SQL query:

```
>>> from author import models
>>> all_authors = models.Author.objects.filter(email__endswith='@
gmail.com').values_list('name').query
SELECT "author_author"."name" FROM "author_author" WHERE "author_
author"."email"::text LIKE %@gmail.com
```

In Django, we have certain queries that are non-QuerySet, such as `count`. When we try to retrieve a query for a non-QuerySet object, we get the following error:

```
>>> from author import models
>>> author_count = models.Author.objects.filter(email='a').count()
Traceback (most recent call last):
  File "<console>", line 1, in <module>
AttributeError: 'int' object has no attribute 'query'
```

To retrieve these queries, we can use the `ConnectionProxy` object by importing it via `django.db import connection`; all database queries are passed through this proxy and are logged. Since the connection object would log all the database queries whenever Django hits the database, it stores all the queries one after another in the form of a list, where the latest database query is appended to the last position of the list. Each connection query has two attributes – the raw SQL query in the `sql` key and the total execution time of the given query in seconds in the `time` key:

```
>>> from django.db import connection
>>> from author import models
>>> author_count = models.Author.objects.filter(email='a').count()
>>> connection.queries[-1]
{'sql': 'SELECT COUNT(*) AS "__count" FROM "author_author" WHERE
"author_author"."email" = \'a\'', 'time': '0.166'}
```

The raw SQL query is highlighted.

> **Important note**
>
> The `ConnectionProxy` object only logs when the Django server is running in `DEBUG=True` mode, which means we would be able to retrieve the queries while working on our local development server. For production, there are other monitoring tools such as `newrelic`, which can help us retrieve information in a much better way. We shall learn more about `newrelic` in *Chapter 14*.

Normalization using Django ORM

Database schema design is a core skill for any developer. There is no right or wrong when we design a database schema – we simply follow the basics and reason out certain decisions. In RDBMS, three types of relationships satisfy all the use cases:

- One-to-one relationships, where each entity has a maximum of one related entity
- One-to-many (foreign key) relationships, where one entity can have multiple related entities (and vice versa)
- Many-to-many relationships, where multiple entities can be related to multiple other related entities

As a rule of thumb, we should always start with a normalized database schema and then slowly migrate to the denormalized schema as per the data access pattern and performance bottleneck. Let's learn how and when to implement normalization in Django.

OneToOneField

As we just mentioned, the one-to-one relationship means that for every entity, there will be only one related entity. For example, each blog would have only one cover image associated with it, so we can use OneToOneField here to establish the relationship:

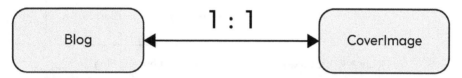

Figure 2.4: A one-to-one relationship between Blog and CoverImage

In Django, OneToOneField is like ForeignKey, with an additional constraint of unique=True on ForeignKey:

```
class Blog(models.Model):
    title = models.CharField(max_length=200)
    content = models.TextField()
    author = models.ForeignKey(Author, related_name='author_blogs',
on_delete=models.PROTECT)

class CoverImage(models.Model):
    image_link = models.URLField()
    blog = models.OneToOneField(Blog, related_name='blog_ci', on_
delete=models.PROTECT)
```

OneToOneField() can be present in either of the tables – that is, Blog or CoverImage. The other table objects can access the related data using a reverse foreign key lookup, the same concept we learned about earlier in this chapter.

The ForeignKey field

ForeignKey is used to represent one-to-many relationships between two different models. For example, a single author can write multiple blogs, as illustrated here:

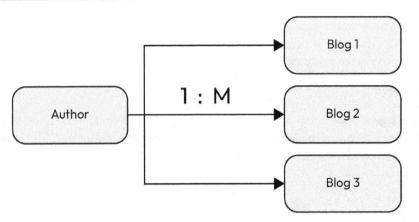

Figure 2.5: ForeignKey relationship between Author and Blog

In Django, we can create a `ForeignKey` relationship between `Author` and `Blog` model:

```
class Author(models.Model):
    name = models.CharField(max_length=100)
    email = models.EmailField()
    bio = models.TextField()

class Blog(models.Model):
    title = models.CharField(max_length=200)
    content = models.TextField()
    author = models.ForeignKey(Author, related_name='author_blogs',
on_delete=models.PROTECT)
```

In the preceding code, each author object can have multiple blogs linked to them, but the opposite is not possible.

> **Important note**
>
> We can create a recursive relationship in a model by establishing a self-referring relationship. This can be done by using the `models.ForeignKey('self', on_delete=models.PROTECT)` constraint.

One of the important attributes we have in both `ForeignKey` and `OneToOneField` is `on_delete`, which determines what we would do when a parent-related object is deleted. There are multiple options that Django provides, but if you are not sure, then always use PROTECT; this stops the delete operation from happening and saves your data from becoming corrupted.

ManyToManyField

ManyToManyField is used to create a relationship where multiple objects of a model are related to multiple objects of a different model. For example, a blog can have multiple tags, while a tag can be associated with multiple blogs:

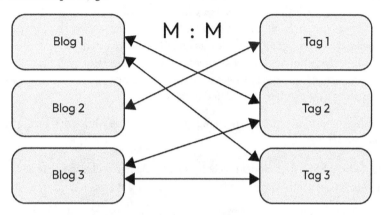

Figure 2.6: ManyToManyField relationship between Blog and Tag

When we create ManyToManyField, Django will create an intermediate table to create the mapping relationship between the two tables by default.

Customizing ManyToManyField

Using through, Django provides an interface that allows us to use any custom model to create mapping relationships between tables. This is an advanced feature and you can learn more from the official documentation: https://docs.djangoproject.com/en/stable/ref/models/fields/#django.db.models.ManyToManyField.through.

Django ORM is powerful enough to provide a lot of advanced database-normalized queries.

Important note

The Django documentation mentions tons of different features we can use while working with relationships in Django. We have only scratched the surface in this section, so check out the official documentation page to learn more about them: https://docs.djangoproject.com/en/stable/ref/models/fields/#module-django.db.models.fields.related.

We used the on_delete option for every normalized field. It is a mandatory option for ForeignKey and OneToOneField. Let's explore the different options for on_delete that are provided by Django.

Exploring on_delete options

In the previous section, while working with normalized data, we saved the data in multiple tables, and we linked each entry to one another using a database field. Since the data is saved in different tables, what would happen if someone deletes a row that has a reference to another table? This is exactly the use case the on_delete keyword solves. For example, we have the Author table and then the Blog table. Each blog is written by an author, so if we have author_1, who has written 10 different blogs, then in our database, the Author table has an entry for author_1, while the Blog table has entries for the 10 different blogs. Now, if someone deletes the author_1 entry, then all the 10 blogs' data will be corrupted. To fix this problem, we can tell Django what to do when someone tries to delete the author_1 entry if there are related entries.

There are seven different options provided by Django that we can use while configuring on_delete while defining our model:

- CASCADE: When the related model object is deleted, Django will also delete all the dependent model objects.

- PROTECT: Django will not allow any model object that has a dependent model object associated with it to be deleted.

- RESTRICT: This prevents the referenced object from being deleted by raising RestrictedError.

- SET_NULL: This will set the foreignkey value to null. It is only possible when null is set to true.

- SET_DEFAULT; This will set the foreignkey value to its default value.

- SET: Use a custom function to generate a custom value for the foreignkey field.

- DO_NOTHING: Take no action.

Now, let's learn about model inheritance and how we should take advantage of inheritance while coding in Django.

Using model inheritance

When implementing any model in Django, we use the concept of standard Python inheritance to inherit the base class from django.db.models.Model. While using model inheritance, Developers can choose whether each parent model will have a database table via **multi-table inheritance**, or whether the parent models will be abstract models without any physical database table via **abstract base classes**. Let's learn about the different types of model inheritance Django provides.

Abstract base classes

When building different application logics in Django, we'll come across models that have almost all the fields and multiple common model methods, leaving just a couple of fields and properties aside. These are the perfect scenarios where we should use abstract models.

For example, it is always good practice to have a timestamp in all the model objects that are created rather than repeating the same code again and again; we can create a base timestamp abstract model and have all the other models inherit it:

```
class BaseTimeStampModel(models.Model):
    created_at = models.DateTimeField(auto_now_add=True)
    updated_at = models.DateTimeField(auto_now=True)

    class Meta:
        abstract = True

class CoverImage(BaseTimeStampModel):
    image_link = models.URLField()
```

When we use this model, only one `CoverImage` model will be created alongside the `created_at` and `updated_at` fields. Note the additional abstract attribute we have added to the `Meta` class: the `abstract` flag defines that the model is abstract, so if we do not add the `abstract` flag, then our migration would create an actual database model table in the database.

Multi-table inheritance

In multi-table inheritance, each model would have a corresponding table in the database. Here, Django automatically creates a `OneToOneField` relationship field under the hood to support multi-table inheritance.

However, you should avoid using multi-table inheritance since each child table gets a new parent table associated with a one-to-one relationship. This causes the unnecessary creation of multiple tables, and since Django creates the relationship under the hood, you have very little control over it. It is always preferred to use an abstract model or custom `OneToOne` relationship. In short, avoid using multi-table inheritance in production.

Proxy model

There are certain times when we have Django models that need to be exposed to different Django applications or only modify the Python behavior of the model class. Proxy models are perfect for such use cases. They can be considered equivalent to the concept of Views in SQL, which are **virtual tables**, as shown here:

```
class Author(models.Model):
    first_name = models.CharField(max_length=100)
    last_name = models.CharField(max_length=100)
```

```
    email = models.EmailField()

class BlogAuthor(Author):
    class Meta:
        proxy = True
    def perform_something(self):
        pass
```

In this example, `BlogAuthor` is a proxy model that won't have any table created in the database but is a wrapper over the `Author` model. Also, note the `proxy` flag, which we have to set in the `Meta` class to enable the proxy behavior of a model.

So far, we have explored the concepts of the Django model, learning how we can create different fields and models in Python. Next, we'll explore how this Python code gets converted into database tables and how Django provides us with the migration interface.

Understanding the crux of Django migrations

Migration is one of the most dreaded words for an engineer, be it migrating an old framework to a new framework, migrating data from an old system to a newer system, or in our case, migrating database schema-related changes. In fact, in a lot of early-stage startups, developers use NoSQL primarily because the business is in such a nascent stage; developers don't want to make database migrations every other day as the business requirements have to adapt to the new database schema requirements.

However, Django solves the database schema migration issue out of the box. In this section, we'll learn how Django handles migration under the hood and how we can work with Django migrations like a pro to solve any challenge thrown at us.

Demystifying migration management commands

Whenever we ask any developer how to perform migration in Django, the most common answer we get is that we can perform migrations by running the following command:

```
python manage.py makemigrations
python manage.py migrate
```

Though this is how Django performs migration, Django abstracts a lot of intricacies under the hood by exposing these two management commands. Let's deep-dive into how and what Django performs under the hood when we run these commands.

The makemigrations command

Once we create a new model or update an existing model, we run the `makemigrations` command to create a new migrations file depending on the previous database schema and the new database schema after the model changes are performed. This is achieved by comparing the previous migrations files present in the app's `migrations` folder; the `migrations` folder will have the first file as `0001_initial.py`. Subsequent files are created using the sequential file numbers.

Django generates the new migration filename automatically using the type of difference it has identified. Django's migrations system can identify the difference between the current `models.py` schema and the existing database schema using the `migrations` folder. To give a meaningful name to the newly created migration file, use `makemigrations -- name`.

We should also remember that the migration files that are generated in Django are not SQL files but Django-Python scripts.

> **Important note**
>
> We should have our `migrations` folder committed to the VCS (such as `git`) so that during collaboration, developers don't face database schema inconsistency. While collaborating, it is common to get duplicate file numbers in the same app in the `migrations` folder. Though Django has us covered in these cases, it is always recommended to have system tests in place that can flag any such conflict so that we can resolve these issues manually after verifying the database-related changes.

Once the migration files have been created, we need to perform the migration to the database using the `migrate` command.

The migrate command

The `migrate` command executes the newly generated migration files and performs the database schema changes. Once the database schema has been updated for a given migration file, it creates a new entry in the `django_migrations` table. This table is used by Django to identify which migration changes have been applied to the database and which are the new ones. For example, *Figure 2.7* shows a screenshot of the `django_migrations` table with three main columns:

- `app`: The app to which the migration file belongs
- `name`: The filename present in the `migrations` folder that got executed
- `applied`: The timestamp of when the migration was applied:

id	app	name
1	contenttypes	0001_initial
2	auth	0001_initial
3	admin	0001_initial
4	admin	0002_logentry_remove_auto_add
5	admin	0003_logentry_add_action_flag_choices
6	contenttypes	0002_remove_content_type_name
7	auth	0002_alter_permission_name_max_length
8	auth	0003_alter_user_email_max_length
9	auth	0004_alter_user_username_opts
10	auth	0005_alter_user_last_login_null
11	auth	0006_require_contenttypes_0002
12	auth	0007_alter_validators_add_error_messages
13	auth	0008_alter_user_username_max_length
14	auth	0009_alter_user_last_name_max_length
15	auth	0010_alter_group_name_max_length

Figure 2.7: The django_migrations table

Apart from these two primary management commands, Django also provides us with two more commands:

- `sqlmigrate`: This gives us the SQL statement for migrations
- `showmigrations`: This shows the current project migration and the status

Now, let's cover a few tricks on how we can use Django migration commands for different use cases.

Performing database migrations like a pro

In this section, we'll learn how to handle migrations and a few things that we should avoid while working with Django migrations.

Perform reverse migrations

Most of the time, we are performing migrations forward, which means that we have made some changes to our models and then run the two magical commands to let Django take care of the database schema change for us. But when we work between features, we must switch between our local branches; though the code will change, our database schema won't change, and that's where we start facing database schema issues. Django has a solution for this problem; just like how we can make new schema changes to the database, Django also has a way to revert the database schema changes that we perform.

For example, let's say that the blog app has four `migrations` files, and the last migration file was created in the current working branch. When we switch the branch to `main`, `main branch` only has three migration changes, so we want to have our three database migrations. To achieve this, we would use Django's reverse migration. By using the Django `migrate` command with the migration file number (`python manage.py migrate <app_name> <migration_num>`), we can perform reverse migration.

Let's check out a real reverse migration example. In this example, we are performing reverse migration and showing the output we would receive in the console:

```
~ python manage.py migrate blog 0001
Operations to perform:
  Target specific migration: 0001_initial, from blog
Running migrations:
  Rendering model states... DONE
  Unapplying blog.0002_coverimage_alter_blog_author... OK
```

The preceding code removed the database changes we had applied in `0002` migrations.

> **Important note**
>
> Sometimes, you may get a reversal failures message in the terminal when you run the reverse migration command. This happens due to data dependencies or another app-related dependency; you must work on them on a case-to-case basis. The solution to such a problem is beyond the scope of this book and unfortunately, there are not many resources to help with this.

Use fake migrations

When we manually apply certain database schema change operations, we can use this powerful fake migration to get our database and Django migration system back in sync. For example, let's say we have to create a new column in a large table, which we know might take a long time to execute; we don't want to perform this operation via the Django management command. Instead, we'll want to ask our DBA to perform the operation. Then, once the new column is added, we can run the `migrate` command with a `fake` flag. By using the Django `migrate` command with a `fake` flag (for example, `python manage.py migrate <app_name> <migration_num> --fake`), we can perform fake migrations.

Let's learn how to run a fake migration. To execute a fake migration, we need new migration file changes that we need to apply. If we update our model code with a few changes, this will create database migrations. Update the author models to add `db_index` to the `name` field and add a new `demo_field`, as shown here:

```
class Author(models.Model):
    name = models.CharField(max_length=100)
    demo_field = models.TextField(default='demo')
    email = models.EmailField(unique=True)
    bio = models.TextField()
```

Now, run the `python manage.py makemigrations` command; we will have a new fake field in the `author/migrations` folder. We do not want to apply the changes to our database, so we will fake the migrations. To fake our migrations, we are going to execute the `migrate` command for `author app 0002` as a fake migration:

```
~ python manage.py migrate author 0002 --fake
Operations to perform:
  Target specific migration: 0002_author_demo_field, from author
Running migrations:
  Applying author.0002_author_demo_field... FAKED
```

Here, the `0002` migration file changes were faked – that is, no database schema change was performed, but we added an entry to our `django_migrations` table anyway.

Now, to revert these fake database migrations changes, we can use the `migrate` command:

```
~ python manage.py migrate author 0001 --fake
Operations to perform:
  Target specific migration: 0001_initial, from author
Running migrations:
  Rendering model states... DONE
  Unapplying author.0002_author_demo_field... FAKED
```

Please remember that if we don't use a `fake` flag while reverting, then we will face an error because the database migrations that we are trying to revert were never actually applied to the database table.

> **Important note**
>
> Please be careful while working with fake migration changes in production. Only apply fake migrations when the database schema is the same as the existing Django model definition.

Avoid custom migration for data migrations

Django provides a way to create custom migration files, and a lot of times, developers are tempted to create custom migration files to perform data migrations. However, whilst this is a great way to do it, I would always avoid performing it this way. Why? Because the same migration file that you will run for data migration on production will also be run on other developers' local systems and generally, data migration files will need data files; this causes an issue since this migration file is compatible with the production environment but not compatible in local and staging applications.

Create a system check on duplicate migrations

You should create system checks that will prevent developers from merging duplicate migration files with the same migration file number. Whenever Django encounters such duplicates, it will give you options to choose from. Though they are perfect solutions, if you have CI/CD in place, then it might be too late for you to get these options.

Adding new fields

Always add the default value while inserting a new column into an existing database table. Prefer to add null values since database migration takes much less time with null values compared to some random values.

Now that we've learned how Django migrations work, let's learn about the best practices we should follow while working with models and ORM.

Exploring best practices for working with models and ORM

Django's official documentation is a very detailed guide on how to work with models and ORM, and it helps you write complicated queries using ORM. But when you create real-world applications, it is more than just writing good queries; there is a lot of nitty-gritty that you should know or get to know after making mistakes yourself. In this section, we are going to learn about all those small tricks that can help us work better with Django models and ORM.

Use base models

Abstract base models are one of the most preferred model inheritances in Django. We should create common base models whenever we find duplicates of the model attributes. One of the most useful base models that I have found is `TimeStampedBaseModel`, which has `created_at` and `updated_at` fields. Using this base model to create any model will automatically add the `created at` and `updated at` timestamp field; for example, whenever someone creates a `DemoModel` object or updates it, we shall have respective timestamps. Let's look at the implementation:

```
from django.db import models

class TimeStampedBaseModel(models.Model):
    created_at = models.DateTimeField(auto_now_add=True)
    updated_at = models.DateTimeField(auto_now=True)

    class Meta:
        abstract = True
class DemoModel(TimeStampedBaseModel):
    name = models.CharField(max_length=100)
```

We should create a common Django app that will have all global base models, such as `TimeStampedBaseModel`; this will let us use `TimeStampedBaseModel` in different Django apps throughout the project.

Use timezone.now() for any DateTime-related data

Whenever you are working with any `DateTime`-related data, always use the `django.utils.timezone.now()` function. The regular Python `datetime.now()` function is not timezone aware, so it will not save the timezone in the timestamp database field. However, `timezone.now()` provides us with timezone information that will help us save the timezone in the database.

> **Important note**
>
> Django settings have a `TIME_ZONE` variable, which is used to define the timezone; by default, every Django project is set to UTC, so you will need to configure the settings to the respective timezone. You can learn more about this here: `https://docs.djangoproject.com/en/stable/topics/i18n/timezones/`.

How to avoid circular dependency in models

Circular dependency is a nightmare for developers. It is common to get circular dependency in mature code bases and when it occurs between models, it becomes difficult to resolve. This happens when we have foreign key references that are interdependent between two apps.

For example, check out the example shown in *Figure 2.8*. We have two app models interdependent of each other, creating a circular dependency error:

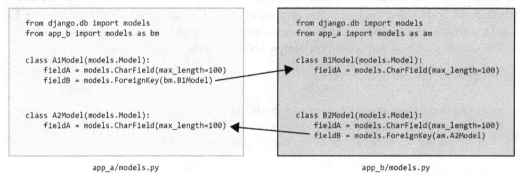

```
from django.db import models
from app_b import models as bm


class A1Model(models.Model):
    fieldA = models.CharField(max_length=100)
    fieldB = models.ForeignKey(bm.B1Model)


class A2Model(models.Model):
    fieldA = models.CharField(max_length=100)
```

```
from django.db import models
from app_a import models as am


class B1Model(models.Model):
    fieldA = models.CharField(max_length=100)


class B2Model(models.Model):
    fieldA = models.CharField(max_length=100)
    fieldB = models.ForeignKey(am.A2Model)
```

app_a/models.py   app_b/models.py

Figure 2.8: When two apps have circular dependency due to model reference

Django gives us a way to avoid this circular dependency error via a string-based foreign key reference. For example, in *Figure 2.9*, `A1Model` will use `app_b.B1Model` in the foreign key reference and we don't have to import the models of `app_b`; similarly, `B2Model` will be updated to have `app_a.A2Model` as a string reference. After we remove imports from the `models.py` files, there won't be any circular dependency error, as shown here:

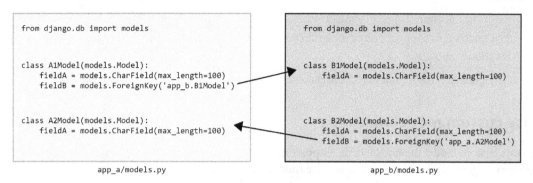

Figure 2.9: Circular dependency resolved using a string-based reference

The string reference of models we add in the `ForeignKey` field is automatically created as a `ForeignKey` reference by Django under the hood.

Define __str__ for all models

During the development phase, it is common to work on the Django shell, so it is very important to identify model objects using human-readable strings rather than random IDs. As an example, for an author object such as `<Author: Author object (1)>`, if we don't define `__str__` in the author model definition, we shall get the output shown in *Figure 2.10*:

```
>>> from author import models
>>> models.Author.objects.create(name='abc', email='a@b.com', bio='xyz')
<Author: Author object (1)>
```

Figure 2.10: The Author object string's output when we don't have the __str__ method defined

Now, let's define the `__str__` method for the `Author` model:

```
class Author(models.Model):
    name = models.CharField(max_length=100)
    email = models.EmailField()
    bio = models.TextField()

    def __str__(self):
        return f'{self.name}-{self.id}'
```

By using this model, if we create a new object, we shall get `Author: <name>-<id>` as output, as shown in *Figure 2.11*, which is much better to read and work with:

```
>>> from author import models
>>> models.Author.objects.create(name='abc', email='a@b.com', bio='xyz')
<Author: abc-2>
```

Figure 2.11: The Author object string's output when we have the __str__ method defined

Use custom model methods

Create custom model methods whenever possible to keep your data interface clean. For example, if we have to get the full name of an author, we should use model methods to avoid repetitive business logic in the application layer, like so:

```
class Author(models.Model):
    first_name = models.CharField(max_length=100)
    last_name = models.CharField(max_length=100)
    email = models.EmailField()
    bio = models.TextField()

    def get_author_name(self):
        return f'{self.first_name} {self.last_name}'

    def get_short_bio(self):
        return f'{self.bio[:200]}...'
```

Here, the get_short_bio and get_author_name model methods help us retrieve information by just calling the object with the method. For example, <authorObj>.get_author_name() would return the full name of the author we have filtered.

> **Important note**
> You shouldn't add too much logic inside the model method as adding all the business logic to the model can cause it to explode and your code will start to smell. An ideal approach would be to keep a balance and try to identify the limit whenever your model starts to become difficult to test.

Note that model methods only work for single objects, not on QuerySet objects; when you find yourself writing similar filter or ORM queries again and again, you should start considering writing a custom model manager.

We won't dive into the details of how to implement model managers since the official documentation already goes into this in detail. You can read more about it here: https://docs.djangoproject.com/en/stable/topics/db/managers/.

Keep the default primary key

By default, Django gives your auto-increment `Integer` field, which is an `id` field, to all the models you define. For most regular use cases, this ID should be kept as your primary key since it gives you a couple of added advantages, such as filtering entities based on the primary key range, sorting queries based on creation (since the primary key is always auto-incremented), and easy human readability of the primary key.

You should have a strong reason to use an ID field other than the regular `Integer` auto-incremented `id` field. Most applications use UUID or some other unique identifier to replace the existing one. Here are some of the reasons you might want to consider UUID:

- The total number of rows is going to outgrow the `Integer` range
- You don't want users to guess the next or previous entries using ID

> **Important note**
> In many data-sensitive applications, it is recommended to use both auto ID and UUID, where the default ID can be used internally and UUID can be exposed to the end user.

Use transactions

By default, Django ORM can commit the data to the database whenever `create()`, `save()`, or `update()` methods are called. When a business use case needs you to update/create multiple rows in a single request, this might become a data corruption problem if transactions are not used.

For example, when a user registers, we have to create a `User` table entry and a `Profile` table entry. However, for some reason, the `User` table entry was created but the backend crashed before the `Profile` table entry was created. Since our application will need both `User` and `Profile` table entries, the user is unable to access our application.

To avoid such scenarios, it is always advisable to wrap queries inside transactions so that whenever there is any issue during the whole process, the whole transaction is rolled back:

```python
from django.db import transaction

def viewfunc(request):
    # This code executes in autocommit mode (Django's default).
    # do_db_stuff()
    with transaction.atomic():
        # This code executes inside a transaction.
        pass
        # do_more_db_stuff()
        # do_one_more_db_stuff()
```

In this example, the `do_db_stuff()` function will perform the default auto-commit on any query written inside the function, while the other two functions will only perform commits after both of them are successful.

> **Read nore**
>
> Django provides a wrapper over the database transaction concept. A few important topics any Django developer should know about regarding Django transactions are the `atomic` transaction decorator, `Savepoints` to manage better commit and rollback control, and database-dependent transaction logic – that is, PostgreSQL and MySQL have different transaction locks. You can read more in Django's official documentation: `https://docs.djangoproject.com/en/stable/topics/db/transactions/`.

Avoid generic foreign keys

In Django, `GenericForeignKey` is a concept where one model is allowed to have a relationship with any other model in the whole project. Though this is one of the unique features of Django ORM and we have seen this as an interview question in a lot of companies, we should avoid using it in production. Since no constraint is added to the table, it can cause data integrity issues and as the data's size increases, it causes queries to become slower and slower. `GenericForeignKey` gives us a faster implementation advantage but in the long run, this becomes a bottleneck.

> **Read more**
>
> The official documentation doesn't mention all the pitfalls and advantages in detail. Luke Plant, one of the core contributors of Django, narrowed down all the pros and cons of `GenericForeignKey` in this blog: `https://lukeplant.me.uk/blog/posts/avoid-django-genericforeignkey/`.

Use finite state machines (FSMs)

Most applications will need us to maintain some state for the data. For example, a blogging application would have multiple stages, such as *submitted > reviewed > published*. The stages should be captured using a database field, `state`. When a new blog is created, the `state` field will have *submitted* as the `state` value. After the blog is thoroughly checked and edited, the state field will have *reviewed* as its value. Finally, when published, the `state` field will have *published* as its value. To programmatically implement this logic in our application, we would have to implement multiple validations and state management.

We can write our own custom logic and validation to handle multiple corner cases, but why should we reinvent the wheel? There is a great package that gives us an interface to just worry about the business logic while it takes care of the underlying infrastructure. I would recommend using Django FSM for any state management-related use case.

> **Read more**
>
> `django-fsm` is a Django package that helps with simple declarative state management for Django models. Django FSM uses decorators to create a cleaner interface and introduces a new `FSMField`, which is a specialized character field for implementing application state management better. You can learn more about Django FSM here: `https://github.com/viewflow/django-fsm`.

Break the model into packages

It is recommended to not have a lot of models inside a single app. However, there can be legacy reasons why new apps were not created and a single model file is 10,000+ lines of code long with 10+ different models. Though there is no such hard limit on how many lines of code a single file can have, it becomes a nightmare for a developer to work on such humongous files. Luckily, Django provides a way to break `models.py` into smaller files and have `models` as a package.

To do so, just create a folder called `models` with an `__init__.py` file, then add all the models you want to add to the package:

```
#### blog_app/models/__init__.py
from .comment import Comment
from .post import Post

#### blog_app/models/comment.py

class Comment(models.Model):
    # comment fields

#### blog_app/models/post.py

class Post(models.Model):
    # post fields
```

This implementation lets us use `models` in the same way as the default single `models.py` file.

> **Important note**
>
> We are not using `from .post import *` to avoid importing unnecessary imports.
>
> You can read more about how Django supports organizing models into a package: `https://docs.djangoproject.com/en/stable/topics/db/models/#organizing-models-in-a-package`.

Now that we've learned a lot of best practices that can help us work with Django ORM and models like a pro, let's learn how to perform optimizations in the data layer using Django.

Learning about performance optimization

In today's modern web world, any API response between 0.1 to 1 second is considered a good overall response time. In the whole request-response cycle, the database often becomes the bottleneck and we see slower responses, primarily due to bad queries or bad database design decisions.

Therefore, in this section, we'll learn how to check for performance optimizations in the Django ORM and database layer. In the previous sections, we learned how to work with models and design database schemas using different normalization techniques, but here, our primary goal is to learn how to use Django ORM efficiently to gain performance optimizations.

One of the first techniques we should understand to gain insight into our application performance is using the `explain` feature of the database query.

Demystifying performance using explain and analyze

As we develop applications in today's world, a lot of people might frown upon us when we say we are using an RDBMS since most junior naïve developers get the notion that SQL is inherently slow. SQL performance depends on what query we have used to access our data and whether the database schema is designed to handle the data access pattern. SQL gives us a powerful built-in tool called `explain` to help us understand how it is going to execute the database query under the hood.

Let's consider the example of you traveling from your current location to a given destination. The first thing you would do is open Google Maps and look for the directions, which would tell you how to travel and how much distance you would travel to reach your destination.

Consider `explain` as Google Maps for your queries. `Explain` will tell you how the query is going to be executed and the different steps it would follow to give you the result of the query. The Django ORM layer provides a wrapper over the `explain` SQL interface, which you can use to get the explain-analyze plan in Django itself:

```
>>> print(Blog.objects.filter(title='My Blog').explain())
    Seq Scan on blog   (cost=0.00..35.50 rows=10 width=12)
      Filter: (title = 'My Blog'::bpchar)
```

Django provides additional support for the `explain` interface for PostgreSQL by passing the `verbose` and `analyze` flags – that is, `explain(verbose=True, analyze=True)`. By passing these flags as `true`, we can get a detailed SQL execution plan analysis of the SQL statement run by Django ORM, which is equivalent to `explain plan`.

> **Important note**
> Discussion regarding how to read the `explain` and `analyze` results of `explain analyze` is beyond the scope of this book since it needs a lot of additional context. You can learn about them in the respective database documentation: `https://www.postgresql.org/docs/current/sql-explain.html`.

Now that we've learned about various data access patterns and how to find bottlenecks, let's consider one of the first solutions that solves most performance issues: adding index to columns.

Using index

index is one of the most widely used performance optimization solutions. Let's try to understand what an index is. Imagine that you are trying to find Mr. A's home in a block of flats. The building may have a flat numbering structure where the 2nd flat on the 14th floor may be called 1402. Similarly, 2609 would mean that you have to go to the 26th floor and find the 9th flat on that floor. If you had the flat number, it would be easy to find Mr. A. But if you only know Mr. A's name and not his flat number, you would have to knock on every door in the building until you find him.

Now, let's consider this in terms of the database: here, the building is a table, the flat is each row, and the flat number is index. If someone tries to find a particular row using index, then they would get the response much faster compared to trying to find a given row using a field that is not indexed. Under the hood, most databases use the **B-tree** as the data structure to store the indexes, which enables us to access any given row in O(logN) time complexity.

Now that we understand the advantage of using index, let's add index to tables using Django models. In Django, we have to use the Meta class to define index for columns. For example, if we find that the Author table has a lot of ORM filters on the first name, then we should add index to the first_name column, which helps us take advantage of index to get a faster response. The following model definition helps us add index to the first_name column of the Author model:

```
class Author(models.Model):
    first_name = models.CharField(max_length=100)
    last_name = models.CharField(max_length=100)
    email = models.EmailField()
    bio = models.TextField()

    class Meta:
        indexes = [
            models.Index(fields=['first_name']),
        ]
```

There are a few important points around indexes:

- By default, if we are using the default id field for any model, then it is already indexed
- Any primary key field, foreign key field, and one-to-one field is indexed by default
- Using index helps us to retrieve the data faster, but it slows down writing data to the given table
- The clustered index should be used after analyzing the data access patterns

Indexes can solve most of our performance issues and should always be used as the first optimization for large tables with high read volumes. Django ORM can also help us gain a significant performance boost if used correctly. In the following section, we'll learn how to use Django ORM like a pro.

Using Django ORM like a pro

Django's official documentation provides a detailed list of how you can use Django ORM to perform any kind of database-related query. It can be overwhelming at times due to the vast list of ORM features available, so in this section, we'll just learn about a small number of Django ORM optimizations you can perform.

> **Important note**
> The idea of this section is to give you a TL;DR version of what you can perform using Django ORM and direct you to the official documentation, where you can learn about it in detail.

Exists versus count

It is common to check if a given entry is present in a database or not. The first query that developers typically choose is a filter query, after which they check the length of the QuerySet or use the `count()` query. While working with a small dataset, it is okay to use a filter or `count()` query; however, when working with large tables, the `count` statement can take a performance hit. For better performance, it is always advised to use the `exists()` query when you want to find out if a particular entry is present in the database or not.

Taking advantage of the lazy loading of QuerySets

QuerySets are lazy-loaded by default, so you should take advantage of this property whenever you are writing code. We can add filter queries back-to-back and Django will not execute them until we evaluate them. In the following example, we're hitting the database with just one query when we call the `print` statement:

```
some_query = Book.objects.filter(author_id=1)
another_query = some_query.filter(name__startswith='a')
yet_another_query = another_query.filter(name__endswith='s')
print(yet_another_query)
```

All three Django ORM filter statements will be executed at once when we run the `print` statement; no database query will be fired before the `print` statement, and we will only hit the database when we execute the last `print` statement. While working on the application logic, we should remember to lazy load QuerySets and not evaluate any query unnecessarily.

> **Important note**
>
> While lazy loading QuerySet helps us save additional database calls when we have application logic in multiple tables, it is important to make sure we are not letting Django create joins on non-indexed fields. For example, if we return a QuerySet from a given function and use the result in another QuerySet, it will automatically create a join between two queries and cause performance degradation. But if we evaluate the QuerySet from the result and use it, then our response will be much faster. This is purely application and uses case-dependent, so developers should keep this in mind while writing queries using Django ORM.

Using select_related and prefetch_related

While creating different business logic, we will come across implementations where we need to access data from different tables. In SQL, we can perform joins to retrieve data from multiple tables in a single query; similarly, Django also provides us with `select_related` to fetch data from multiple tables by using the join query under the hood.

Let's create a database debug decorator to retrieve a few query stats. The `database_debug` decorator intercepts all the queries that are executed by a single function, gets the total number of queries hitting the database, and formats all the SQL queries in a readable format:

```
from django.db import connection
from django.db import reset_queries

def database_debug(func):
    def inner_func(*args, **kwargs):
        reset_queries()
        results = func(*args, **kwargs)
        query_info = connection.queries
        print(f'function_name: {func.__name__}')
        print(f'query_count: {len(query_info)}')
        queries = [f'{ query["sql"]}\n' for query in query_info]
        print(f'queries: \n{"".join(queries)}')
        return results
    return inner_func
```

To understand the performance implication when we are not using `select_related` and when we are using it, let's check out this simple model for `Author` and `Blog`:

```
from django.db import models

class Author(models.Model):
    name = models.CharField(max_length=100)
    email = models.EmailField()
    bio = models.TextField()
```

```
class Blog(models.Model):
    title = models.CharField(max_length=200)
    content = models.TextField()
    author = models.ForeignKey(Author, on_delete=models.PROTECT)
```

Now, we'll fetch all the first names of the authors for a list of blogs:

```
@database_debug
def regular_query():
    blogs = models.Blog.objects.all()
    return [blog.author.name for blog in blogs]
## OUTPUT
function_name: regular_query
query_count: 4
queries:
SELECT "blog_blog"."id", "blog_blog"."title", "blog_blog"."content",
"blog_blog"."author_id", "blog_blog"."created_at", "blog_
blog"."updated_at" FROM "blog_blog"
SELECT "author_author"."id", "author_author"."name", "author_
author"."email", "author_author"."bio" FROM "author_author" WHERE
"author_author"."id" = 1 LIMIT 21
SELECT "author_author"."id", "author_author"."name", "author_
author"."email", "author_author"."bio" FROM "author_author" WHERE
"author_author"."id" = 2 LIMIT 21
SELECT "author_author"."id", "author_author"."name", "author_
author"."email", "author_author"."bio" FROM "author_author" WHERE
"author_author"."id" = 2 LIMIT 21
```

We have a total of three blogs in the table, and whenever we try to fetch the author's first name for any blog, Django makes an additional query. Considering this example, Django makes one query to fetch all the blogs and then the remaining three queries to fetch the name of the author, which results in a total of four queries. This clearly shows how inefficient this kind of query is in production.

So, let's see how we can improvise by using the select_related feature of Django:

```
@database_debug
def select_related_query():
    blogs = models.Blog.objects.select_related('author').all()
    return [blog.author.name for blog in blogs]

## Result
function_name: select_related_query
query_count: 1
queries:
SELECT "blog_blog"."id", "blog_blog"."title", "blog_blog"."content",
"blog_blog"."author_id", "blog_blog"."created_at", "blog_
```

```
blog"."updated_at", "author_author"."id", "author_author"."name",
"author_author"."email", "author_author"."bio" FROM "blog_blog"
INNER JOIN "author_author" ON ("blog_blog"."author_id" = "author_
author"."id")
```

In this example, we can see that the same functionality can be achieved using a single query (compared to the four queries previously). As per the query, we can see that Django uses the INNER JOIN SQL query to fetch the author model data in a single query.

Django provides another approach to prepopulate data via the prefetch_related syntax:

```
@database_debug
def prefetch_related_query():
    blogs = models.Blog.objects.prefetch_related('author').all()
    return [blog.author.name for blog in blogs]
## Result
function_name: prefetch_related_query
query_count: 2
queries:
SELECT "blog_blog"."id", "blog_blog"."title", "blog_blog"."content",
"blog_blog"."author_id", "blog_blog"."created_at", "blog_
blog"."updated_at" FROM "blog_blog"
SELECT "author_author"."id", "author_author"."name", "author_
author"."email", "author_author"."bio" FROM "author_author" WHERE
"author_author"."id" IN (1, 2)
```

In the preceding code for prefetch_related, Django performs the join operation in memory inside Python, but in select_related, Django performs the SQL join operation.

Here, we have made two queries: the first query gets information about the blog, while the second query retrieves author information for all the authors present in the selected blog, which we got from the first query. Once we have both SQL queries retrieve the information, Django performs the join operation inside Python.

Now that we have a basic understanding of how select_related and prefetch_related work, let's learn a few important points about them:

- select_related uses a single query to fetch all the information and works only for OneToOneField and ForeignKey fields.

- prefetch_related uses two queries to fetch all the required information and works best for ManyToManyField operations and reverse ForeignKey lookups.

- select_related uses a SQL join operation while prefetch_related uses the join operation in Python.

- prefetch_related can also be customized to fetch data from another field. The prefetch() object is used to control the operation of prefetch_related.

> **Important note**
>
> We won't discuss `prefetch` in detail since the official documentation provides an elaborate explanation of it and its use cases: `https://docs.djangoproject.com/en/stable/ref/models/querysets/#prefetch-objects`.

Avoiding bulk_create and bulk_update

While creating/updating multiple entries for a given table, you might be tempted to use the `bulk_create()` or `bulk_update()` method in the name of performance optimizations. These methods are good to use when you have a basic application, but as your project grows, you will find that more complexities are added to the project. The model's `save()` method is not called, so `pre_save` and `post_save` signals will not get triggered. This can cause unexpected behavior in the application.

> **Read more**
>
> To learn more about `bulk_create`, go to `https://docs.djangoproject.com/en/stable/ref/models/querysets/#bulk-create`.

Using get_or_create and update_or_create

`get_or_create` and `update_or_create` help us create an idempotent API interface that can handle concurrent requests without the need for us to create duplicate entries. For example, if the client tries to create a new entry with the same name, then our code won't create duplicate entries and will have just one entry. Note that the unique constraint should be maintained in the database layer and not left in the application logic.

> **Read more**
>
> To learn more about `bulk_update`, go to `https://docs.djangoproject.com/en/stable/ref/models/querysets/#get-or-create`.

Database connection configuration

Using proper database connection configuration in production is very important for a stable and scalable web application. In this section, we'll cover a couple of configurations that you should use in production for better performance.

Using CONN_MAX_AGE

Each Django server process will connect to the database using an individual TCP connection. By default, Django creates a new connection on every request and closes it, after the response is returned. This default behavior might cause you some performance degradation if your application has consistent traffic. Django gives us a way to configure settings to have a consistent TCP connection between our database and application server. This can be configured by adding a suitable value to CONN_MAX_AGE. By default, it is set to 0, which means it will create a new connection on every request. By setting it to a *positive integer value*, the connection will be persistent for the given number of seconds. To keep the connection persistent for an unlimited time, set the value to None. One important factor we need to consider before modifying these settings is that the database must support these many connections; otherwise, we will get an error stating FATAL: too many connections.

> **Read more**
>
> To learn more about configuring Django for persistent connections, go to https://docs.djangoproject.com/en/stable/ref/databases/#persistent-connections.

Using connect_timeout

There can be scenarios where a developer might have written a bad ORM query that is taking a lot of time to execute on the database. During that time, our Django process will be blocked. So, we should configure our database so that it has a timeout for SQL queries that are fired from the Django server so that our whole application doesn't see higher latency due to one request hitting a bad query.

In PostgreSQL, this configuration can be done using the following code (you need to update the highlighted section of the code for other databases):

```python
from django.db.backends.signals import connection_created
from django.dispatch import receiver


@receiver(connection_created)
def setup_query_timeout(connection, **kwargs):
    # Set Timeout for every statement as 60 seconds.
    with connection.cursor() as cursor:
        cursor.execute("set statement_timeout to 60000;")
```

In the preceding code, whenever a connection is established, Django will send a signal to this snippet, which will then fire a query, setting the timeout to 60 seconds for every query sent to the database by the Django server.

Django 3.1 introduced asynchronous behavior to Django Views by natively supporting the asynchronous behavior in the request-response cycle. However, it still had ugly syntax when we had to work with ORMs, as shown here:

```python
from asgiref.sync import sync_to_async


def _get_blog(pk):
    return Blog.objects.select_related('author').get(pk=pk)


get_blog = sync_to_async(_get_blog, thread_sensitive=True)
```

Not many developers adapted to this syntax due to the additional wrapper, but finally, in Django 4.1, we have native support for Async behavior to Django ORM. Let's learn how we can take advantage of Django Async ORM.

Exploring Django Async ORM

Django Async ORM was introduced in Django 4.1 to help developers take advantage of native support of asynchronous behavior in Python. When we use asynchronous Views or code in Django, we cannot use the native synchronous ORM. Using synchronous Django ORM would mean we are calling blocking synchronous code from asynchronous code, which will block the event loop. Django has a way to stop this, and we'll see a `SynchronousOnlyOperation` error thrown by Django during such an operation.

Read more

If you are not familiar with async-await support in Python, please check out the official documentation to understand how Python natively supports async-await: `https://docs.python.org/3/library/asyncio-task.html`.

We should use Django Async ORM with Django Async view/code. The asynchronous query API would give the same response as the original synchronous Django ORM query API. The only difference is how we write the syntax of the query.

For example, let's say we have to find a blog using a `get` query. Let's see what our code looks like when using synchronous code:

```
def get_blog(request):
    blog_details = Blog.objects.get(id=10)
    return JsonResponse({'title': blog_details.title})
```

Now, if we want to use the same `get` operation using asynchronous code, the syntax changes. We have to add the `async-await` keyword and add a before the Django synchronous interface:

```
async def get_blog(request):
    blog_details = await Blog.objects.aget(id=10)
    return JsonResponse({'title': blog_details.title})
```

Django's asynchronous feature is still in active development, and we hope to see more asynchronous support available for the framework. And with this, we have learned how simple the syntax change is for Django Async ORM.

> **Important note**
>
> It is advisable to avoid using Django Async ORM in production unless you have a very specific requirement that requires the use of Async ORM, primarily since the official documentation is also under development without many examples available. For more details, check out the official documentation: `https://docs.djangoproject.com/en/stable/topics/db/queries/#asynchronous-queries`.
>
> At the time of writing of this book, Django 5 transactions are not supported via Async ORM. However, this might change by the time you are reading this section. Please visit `https://docs.djangoproject.com/en/stable/topics/db/queries/#transactions` to check out the latest status.

Now that we have learned how we can use Django Async ORM in our Django project, that brings us to the end of *Chapter 2*.

Summary

In this chapter, we learned a lot of concepts regarding how to work with models and databases. First, we learned how to configure our Django project with ElephantSQL, which enabled us to work with PostgreSQL on our local system without the hassle of setting up a PostgreSQL server. Then, we learned about the basic concepts of ORM and how we can use normalization in Django models, along with the model's inheritance concept.

Migration is a crucial part of any application, and we saw how Django eases the pain of database schema migration. We learned the tricks of using Django migration in our regular development cycle and the best practices we should follow while working with it. Django ORM is one of the pioneers of how frameworks should handle database interaction. After, we covered the best practices that we can follow with Django ORM before looking at a couple of performance optimization tricks that developers can use to build a scalable application.

Django 4.1 introduced the much-awaited Async support in ORM, which helps developers start building asynchronous applications. Overall, in this chapter, we learned all the concepts related to Django and nothing related to DRF. So, in the next chapter, we'll learn how we can connect Django models with DRF serializers.

3

Serializing Data with DRF

In *Chapter 2*, we learned how to work with databases and models using Django. All the concepts that we have learned are related to Django and how developers can make the best use of the Django framework to interact with a database. In this chapter, we shall take another step forward to learn how DRF integrates with Django ORM and models to help us create APIs that will be consumed by client applications.

We shall learn in detail the concepts of serialization and deserialization, and how the client interacts with server applications using the JSON data type. Validating every input and fetching data from multiple tables is quite common for developers, so DRF helps to implement a leaner interface for such use cases, with the help of Serializers. DRF provides Serializers that can be customized to work with Django models out of the box. Finally, we shall learn how we can use Serializers in *APIViews* and *Generic Views*, which we learned about in the first chapter.

We shall cover the following topics in this chapter:

- Understanding the basics of DRF Serializers
- Using model Serializers
- Implementing Serializer relations
- Validating data using Serializers
- Mastering DRF Serializers
- Using Serializers with DRF views

Technical requirements

In this chapter, we shall learn in depth about DRF serializers and how developers can make the best use of serializers in the application code. To follow along with the concepts covered in this chapter, you should have a clear understanding of Django models and ORM. We shall discuss related fields and views, so you should have understood the previous two chapters of the book. Although we shall cover all the concepts of the serializers, it would be helpful to have some understanding of how to create DRF serializers.

Here is the GitHub repository that has all the code and instructions for this chapter: `https://github.com/PacktPublishing/Django-in-Production/tree/main/Chapter03`.

Understanding the basics of DRF Serializers

In the previous chapter, we learned how to use databases using Django models and ORM. However, in the world of web applications, the interaction between the client and server happens using REST APIs, with the JSON request-response format. Let us see how a client-server API interaction looks.

A client application uses the JSON data format to send information to the server via a ReST API. A Django server running on DRF uses, by default, **JSONParser** to parse the data from the JSON format to the Python native format. Then, data in the Python native format would be passed to a serializer to save the data in the Database, using the Django ORM. Similarly, when data is retrieved from the database using the Django ORM, it will be passed through a Serializer to deserialize the data into the Python native format. The Python native format data would be passed through **JSONRenderer** to render the data into the JSON format, finally sending a response to the client.

Figure 3.1 depicts the data flow we discussed, offering a clear understanding of all the underlying data transformations performed by DRF and Django for a client-server interaction.

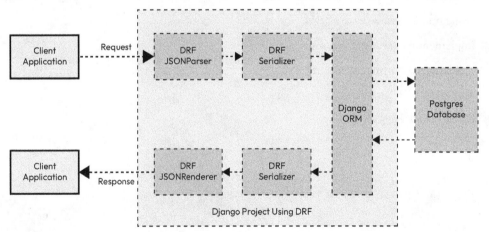

Figure 3.1: The data workflow of the DRF JSONParser, Serializer, and JSONRenderer

---

Configuring custom Parsers and Renderers

Generally, most modern client applications use the JSON content type to interact with servers. If you find that the data sent by the client is not getting parsed by the DRF views, then you should investigate the `Content-Type` header that the client is sent. DRF gives you the flexibility to add different Parsers, and a Parser is picked using the `Content-Type` header defined. Similarly, DRF gives you the flexibility to configure Renderers. You can configure a Parser and Renderer by defining the `parser_classes` and `renderer_classes` attributes in the `APIView` classes, respectively.

DRF Serializers help us to convert complex data types such as QuerySets and model instances into native Python datatypes that can easily be transformed, using the Renderer to return responses to the client. DRF Serializers also provide a deserialization interface that helps us to convert the Python native datatype to complex data types, along with data validation support.

In this chapter, we shall not discuss the generic Serializer class; however, since most application developers use the `ModelSerializer` class to implement business logic, we will discuss that next.

Using Model Serializers

The most common use case for Serializer is to save or retrieve data from Django models. DRF provides `ModelSerializer`, which helps us to integrate DRF serializers to a Django model that can serialize/deserialize fields present in it. `ModelSerializers` maps between the serializable fields and database fields present in the model definition.

Let's work with a few basic `ModelSerializer` examples to understand the concept. We'll use the same `Blog` Django model we built in *Chapter 2* and create `BlogSerializer` to serialize the data. We can define our serializers in any place, but to have maintainable code, it is advisable to have an individual `serializers.py` file for each app that contains all the Serializer definitions:

```
# blog/models.py
class Blog(BaseTimeStampModel):
    title = models.CharField(max_length=200)
    content = models.TextField()
    author = models.ForeignKey(Author, related_name='author_blogs',
on_delete=models.PROTECT)
# blog/serializers.py

class BlogSerializer(serializers.ModelSerializer):
    class Meta:
        model = Blog
        fields = '__all__'
```

Now, let's open the Django shell to explore `BlogSerializer`. First, we'll find all the serializer fields that got linked to our serializer automatically. In the following code snippet, we can see that all the fields that are present in the `Blog` model are added to our serializer:

```
>>> from blog import serializers
>>> print(serializers.BlogSerializer())
BlogSerializer():
    id = IntegerField(label='ID', read_only=True)
    title = CharField(max_length=100,
validators=[<UniqueValidator(queryset=Blog.objects.all())>])
    content = CharField(style={'base_template': 'textarea.html'})
    created_at = DateTimeField(read_only=True)
    updated_at = DateTimeField(read_only=True)
    author = PrimaryKeyRelatedField(queryset=Author.objects.all())
```

Serializers are used to create, update, and retrieve information from the Django models. Let's check the basics of how we can perform all three operations using `BlogSerializer`.

Creating a new model object

Model Serializers improve the overall experience of creating new entries in the Django model via the REST API. As `ModelSerializer` links the fields defined in the Django models to the Serializers, validation of input data is automatically taken care of by the DRF serializer.

For example, in the following code snippet, we can see how we pass the input data to the Serializer. The validators associated with the serializer are called with `<serialize instance>.is_valid()` to validate the input value, returning `True/False`. To save the data to a database, we need to call the `<serialize instance>.save()` method (we should always remember to call the `save()` method only after the `is_valid()` method is called). The validation errors can be retrieved from the `<serialize instance>.errors` attribute.

Let's now check out an example of how we can save a new blog object using DRF serializers. Note that the `author` value of 1 is the primary key of the `Author` object we want to link to the model:

```
>>> from blog import serializers
>>> input_data = {
...     'title': 'new blog title',
...     'content': 'this is content',
...     'author': 1
... }
>>> new_blog = serializers.BlogSerializer(data=input_data)
>>> new_blog.is_valid()
True
>>> new_blog.save()
<Blog: Blog object (11)>
```

DRF also gives us the provision to modify the default `create` behavior. By overriding the `create` method of `ModelSerializer`, we can have custom behavior when we call the `save()` method. For example, let's add a `print` statement to any new object creation call:

```
class BlogCustomSerializer(serializers.ModelSerializer):
    def create(self, validated_data):
        print('*** Custom Create method ****')
        return super(BlogCustomSerializer, self).create(validated_
data)


    class Meta:
        model = Blog
        fields = '__all__'
```

If we want to create multiple objects, we can pass the list of new objects and an additional flag, many, that tells the serializer that multiple objects are going to be created. The following code snippet shows the implementation of creating multiple objects:

```
>>> from blog import serializers
>>> input_data = {
...     'title': 'new blog title',
...     'content': 'this is content',
...     'author': 1
... }
>>> new_blog = serializers.BlogSerializer(data=[input_data, input_
data], many=True)
>>> new_blog.is_valid()
True
>>> new_blog.save()
*** Custom Create method ****
*** Custom Create method ****
[<Blog: Blog object (12)>, <Blog: Blog object (13)>]
```

Once we have created a new entry, we might want to modify the existing value; DRF provides a clean interface to update existing database entries using Serializer.

Updating existing model Objects

To update a model instance with serializer, we have to pass the existing blog object that we want to update as the `instance` property, along with a `partial` flag if we want to update only a specific field. The following code snippet shows an example of how we can update only the title of an existing model instance:

```
>>> from blog import serializers
>>> update_input_data = {
```

```
...      'title': 'new blog title',
... }
>>> existing_blog = models.Blog.objects.get(id=10)
>>> new_blog = serializers.BlogSerializer(instance=existing_blog,
data=update_input_data, partial=True)
>>> new_blog.is_valid()
True
>>> new_blog.save()
<Blog: Blog object (10)>
```

We can then override the `update` method to add custom logic to any update operation used by the Serializer:

```
class BlogCustom2Serializer(serializers.ModelSerializer):
    def update(self, instance, validated_data):
        print('*** Custom Update method ****')
        return super(BlogCustom2Serializer, self).update(instance,
validated_data)

    class Meta:
        model = Blog
        fields = '__all__'
```

The highlighted section in the preceding code snippet shows how we have overridden the existing update method and added a `print` statement.

Now, let's explore how we can use Model Serializers to serialize retrieved QuerySet information from the database.

Retrieving data from the Model object instance

When we pass a model object instance to a Model Serializer, it serializes the data into a dictionary format as per the field definition. In the following code snippet, we pass a single database object to the model serializer class to serialize the values. `blog_obj.data` has the serialized information from the `Blog` object instance:

```
>>> from blog import serializers
>>> from blog import models
>>> from pprint import pprint
>>> tenth_blog = models.Blog.objects.get(id=10)
>>> blog_obj = serializers.BlogSerializer(instance=tenth_blog)
>>> pprint(blog_obj.data)
{'author': 1,
 'content': 'this is content',
 'created_at': '2023-10-01T10:12:09.192181Z',
```

```
 'id': 10,
 'title': 'updated_blog title',
 'updated_at': '2023-10-02T08:21:01.422111Z}
```

Similarly, when we pass a QuerySet to the serializer, we can serialize all the model objects. We need to pass the many flag to the Serializer while instantiating it so that it recognizes that we are trying to serialize multiple model objects:

```
>>> from pprint import pprint
>>> from blog import models, serializers
>>> multiple_blogs = models.Blog.objects.all()
>>> blog_objs = serializers.BlogSerializer(instance=multiple_blogs,
many=True)
>>> pprint(blog_objs.data)
```

Now that we have seen how we can create, update, and retrieve data using Serializers, let's explore how we can configure Model Serializers using the Meta class.

Exploring the Meta class

DRF Serializers follow the concept of Django models to define metadata. Anything that is not related to the field definition of the serializer should be defined inside the inner Meta class.

Let's learn a few important attributes to declare on the inner Meta class for ModelSerializer:

- model: ModelSerializer needs a Django Model mapped to it; to do so, we use the model attribute to map the corresponding Django model to the serializer:

```
class BlogSerializer(serializers.ModelSerializer):
    class Meta:
        model = Blog
        fields = '__all__'
```

In the example, we have a Django model, Blog, that is linked to BlogSerializer using the model property.

- fields: When we declare a Django model to ModelSerializer, we must also define the corresponding fields we want the serializer to use. The fields attribute takes the list of field names as input that we want the serializer to include. It also takes a special value, '__all__', which will indicate that all the fields from the linked model should be used. For example, if we use the following snippet, BlogSerializer will have all the fields that are defined in the Blog model:

```
class BlogSerializer(serializers.ModelSerializer):
    class Meta:
        model = Blog
        fields = '__all__'
```

Now, if we declare fields with a list of model field names, we will be able to see only those values in the serializer output. For example, in the following snippet, we would have only three fields that are defined:

```python
class BlogSerializer(serializers.ModelSerializer):
    class Meta:
        model = Blog
        fields = ['id', 'title', 'created_at']
```

- exclude: In certain cases, we may want to add all the fields to the serializer excluding a few; that is when the exclude property is super helpful. When we define an exclude property, all the fields excluding the ones defined in the list are ignored by the serializer. For example, in the following snippet, we would have all the fields present in the Blog model, apart from updated_at, which is defined in the exclude attribute:

```python
class BlogSerializer(serializers.ModelSerializer):
    class Meta:
        model = Blog
        exclude = ['updated_at']
```

We should always remember that for ModelSerializer, it is mandatory to provide either fields or exclude. Also, the exclude attribute will only map to model fields defined in the Django model class.

- read_only_fields: A single ModelSerializer can be used to both save and retrieve data from a database. Hence, we will have use cases where we want certain fields to be only serialized for retrieving data and not during the write operation. In this example, the updated_at field would have only been serialized when we retrieved data from the database:

```python
class BlogSerializer(serializers.ModelSerializer):
    class Meta:
        model = Blog
        fields = '__all__'
        read_only_fields = ['updated_at']
```

There is another way to define the read-only field – that is while defining the fields. The Serializer field definition takes an argument, read_only, while declaring the field definition. By default, any AutoField is set to read-only behavior.

- extra_kwargs: This attribute is one of the most confusing yet powerful features that a DRF ModelSerializer provides. Since the documentation of DRF doesn't have any details, it becomes quite tricky to understand how to work with extra_kwargs, but we shall check a few of the basic options that can be configured using the extra_kwargs dictionary:

  - write_only: We can define whether a field should be a write-only, Boolean-type value

  - min_length: The minimum length for the field, applicable for string fields

- `max_length`: The maximum length for the given field, applicable for string fields

- `Required`: A Boolean value used to mark the field if it is required

- `Validators`: Used to set a custom validator for a given field

Apart from the ones just mentioned, there are other attributes such as `default` and `source` that we can configure, depending on our use case. Let's check out a basic example of how we can use `extra_kwargs` in a model serializer:

```
class BlogSerializer(serializers.ModelSerializer):
    class Meta:
        model = Blog
        fields = '__all__'
        extra_kwargs = {
            'updated_at': {'write_only': True},
            'title': {'min_length': 100, 'required': True},
        }
```

Each parent key will be the field name, and the properties should be the value of the key – for example, `'updated_at': {'write_only': True}` means the `updated_at` field is write-only.

> **Important note**
> If a field is explicitly defined in the serializer class, then the property defined in the `extra_kwargs` option will always be ignored.

- `Depth`: Django models have foreign key relation fields and other related fields linked. By default, DRF will not retrieve any of the related field values apart from the foreign key. When we define the `depth` attribute, it automatically traverses the relationships to retrieve all the information and, finally, represents it in a JSON structure. The following snippet will retrieve all the foreign key relationship values and return the first level of information:

```
class BlogSerializer(serializers.ModelSerializer):
    class Meta:
        model = Blog
        fields = '__all__'
        depth = 1
```

`depth` is an integer value to determine the depth of relation that the serializer should retrieve, and by default, `depth` is set to 0.

Now that we have learned how we can configure Model Serializers, let's learn another important aspect of Serializers – relations.

Implementing Serializer relations

Django models have three types of relational fields – ForeignKey, ManyToManyField, and OneToOneField (all of them support reverse relationships too). So far, we have learned how to use model serializers efficiently, but let's now explore how we can work with related fields in Serializers.

By default, when a model has any related field, ModelSerializer will link the associated field to PrimaryKeyRelatedField in the serializer definition. A DRF serializer will use PrimaryKeyRelatedField for all kinds of related field serializers, with different options passed.

Here, we will learn how to create related field serializers:

- For ForeignKey, the DRF model serializer assigns PrimaryKeyRelatedField to the serializer

- For OneToOneField, the DRF model serializer assigns PrimaryKeyRelatedField to the serializer with an additional UniqueValidator attached

- For ManyToManyField, the DRF model serializer assigns PrimaryKeyRelatedField to the serializer with two additional flags, many=True and allow_empty=True

By default, PrimaryKeyRelatedField is a read-write field unless we specifically mention the read_only=True flag. Here is an example of how we use PrimaryKeyRelatedField for the corresponding related model fields:

```python
# blog/models.py
class Blog(models.Model):
    author = models.ForeignKey(
        Author, on_delete=models.PROTECT
    )
    cover_image = models.OneToOneField(
        CoverImage, on_delete=models.PROTECT
    )
    tags = models.ManyToManyField(Tags)


# blog/serializers.py
class BlogCustom3Serializer(serializers.ModelSerializer):
    author = serializers.PrimaryKeyRelatedField(
        queryset=Author.objects.all()
    )
    tags = serializers.PrimaryKeyRelatedField(
        queryset=Tags.objects.all(), many=True, allow_empty=True
    )
    cover_image = serializers.PrimaryKeyRelatedField(
        queryset=CoverImage.objects.
all(),      validators=[validators.UniqueValidator(CoverImage.objects.
```

```
all())]
    )

    class Meta:
        model = Blog
        fields = '__all__'
```

In this example, we manually define all the different types of related field serializers that DRF automatically creates under the hood.

What is the QuerySet argument in PrimaryKeyRelatedField?

The "queryset" argument is used to validate field inputs when we try to create a new entry. For read-only operations, "queryset" is not used; hence, it can be ignored when we create serializers that have read-only related fields. It is mandatory to provide a "queryset" argument for any write operation, since it will perform validation using the provided queryset.

DRF provides the flexibility to create nested serializers, which are helpful when we want to fetch nested object data. We can use the depth (which we learned about earlier) configuration to get default fields, but if we need custom fields from nested objects, then nested serializers come in handy.

Working with nested Serializers

We should use nested Serializers when we want to fetch selected fields from the nested objects. For example, let's fetch the author name and author bio for all the blogs. BASerializer exposes only the name of the author and bio, so the final serialized data will only have the name of the author and bio for each blog. Then, the source attribute is used to tell the serializer which database model field it should use to pass the related object information:

```
class BASerializer(serializers.ModelSerializer):
    class Meta:
        model = Author
        fields = ['name', 'bio']


class BlogCustom4Serializer(serializers.ModelSerializer):
    author_details = BASerializer(source='author')

    class Meta:
        model = Blog
        fields = '__all__'
```

We can also pass ManyToManyFields or reverse relationships using nested serializers, by adding the many=True flag. By default, nested serializers are always read-only. To use them for write operations, we must update the create and/or update methods.

While working with related fields, we might have to fetch field values from multiple depths of related tables, and we might not want to use too many nested serializers. DRF provides two different options, using `source` or `SerializerMethodField`, to solve this use case. Let us look at how we can use the `source` argument.

Exploring source

Internally, DRF can scan through nested objects and retrieve the relevant field. The `source` attribute can be considered as a way to map a particular serializer field and database field. We can perform multiple operations using the `source` attribute to do the following:

- Rename a given field for API without making a change to the database model
- Add a serializer read-only field that would access model methods under the hood
- Access multiple levels of an object without writing custom code

Let's explore these operations with examples to get a better understanding. In the following example, `AuthorSerializer` implements three custom fields using the `source` attribute:

```python
# author/models.py
class Author(models.Model):
    bio = models.TextField()

    def fetch_short_bio(self):
        return self.bio[:100]

# author/serializers.py
class AuthorSerializer(serializers.ModelSerializer):
    long_bio = serializers.CharField(source='bio')
    short_bio = serializers.CharField(source='fetch_short_bio')

    class Meta:
        model = Author
        fields = '__all__'
        exclude = ['bio', 'email']
```

Here, we use the `source` attribute to rename the `bio` model field `long_bio`. Then, in the `short_bio` field, we use the `source` attribute to access the model method from the `Author` model, `fetch_short_bio`.

The `source` attribute is powerful enough to access different properties present in the corresponding models. If we want to perform data manipulation, then DRF provides `SerializerMethodField`, which has access to the model object instance to perform any kind of manipulation.

Exploring SerializerMethodField

`SerializerMethodField` is a read-only field that can be defined in any serializer class. It automatically passes the model instance object to the method. For example, if we want to compute the total words in a particular blog, we can use `SerializerMethodField`. In that case, the word_count field is a method serializer field, and by default, DRF looks for `get_<field name>` to get the corresponding method:

```
class BlogCustom5Serializer(serializers.ModelSerializer):
    word_count = serializers.SerializerMethodField()

    def get_word_count(self, obj):
        return len(obj.content.split())

    class Meta:
        model = models.Blog
        fields = '__all__'
```

If we want to use a custom method name, then we can pass the `method_name` argument to the definition. For example, we can name the method `use_custom_word_count` and pass it to `method_name` to link it:

```
class BlogCustom6CustomSerializer(serializers.ModelSerializer):
    word_count = serializers.SerializerMethodField(
        method_name='use_custom_word_count'
    )

    def use_custom_word_count(self, obj):
        return len(obj.content.split())
    class Meta:
        model = models.Blog
        fields = '__all__'
```

We can perform any type of operation in the method – for example, access foreign key field data or make an additional query.

> **Important note**
>
> By default, reverse foreign key relationships are not added in the serializer, but we can manually add them by using the source keyword. Also, be very careful while using a related fields serializer in read operations. Since, by default, there is no select_related or prefetch_related query used, developers can encounter an **N+1 query problem** that causes degradation of performance. We shall discuss how to overcome this problem in the latter part of this chapter, *Avoiding the N+1 query Problem*.

Now that we have seen how we can use related field serializers, we will see how DRF serializers give the validator an interface and implement different types of validation using DRF serializers.

Validating data with serializers

One of the most loved features of DRF serializers is the data validation framework it provides out of the box. By default, whenever a model is defined to the model serializer, it automatically picks the field types and the database field validators, applying them to the serializer. When we pass raw data to the serializer and call the is_valid() method, it evaluates the data against the autogenerated validators and gives us a Boolean value.

Now, let's investigate a few in-depth concepts and the implementation of validators in DRF Serializers.

Customizing field-level validation

DRF serializer fields can have custom validation logic linked to a particular field. By defining a validate_<field name> method in the serializer class, DRF automatically executes the validator whenever the is_valid() method is called. For example, if we don't want to have an underscore inside a title, we can use the following snippet:

```
class BlogCustom7Serializer(serializers.ModelSerializer):
    def validate_title(self, value):
        print('validate_title method')
        if '_' in value:
            raise serializers.ValidationError('illegal char')
        return value

    class Meta:
        model = models.Blog
        fields = '__all__'
```

Here, validate_title would be executed automatically by the DRF serializer whenever we call the is_valid method.

Whenever we create a custom validator, remember to use raise ValidationError on any invalid state. DRF looks for ValidationError particularly.

Defining a custom field-level validator

There are certain fields that are again and again added to different serializers; hence, the same validation logic will be needed. To avoid duplication of the same logic, we can define a validator function and pass it to the `validator` attribute, using `extra_kwargs` or the `field` attribute. In the following example, both the title and content fields have the same validator that checks whether there is an invalid character in the value:

```python
def demo_func_validator(attr):
    print('func val')
    if '_' in attr:
        raise serializers.ValidationError('invalid char')
    return attr


class BlogCustom8Serializer(serializers.ModelSerializer):
    class Meta:
        model = models.Blog
        fields = '__all__'
        extra_kwargs = {
            'title': {
                'validators': [demo_func_validator]
            },
            'content': {
                'validators': [demo_func_validator]
            }
        }
```

Personally, I prefer using `extra_kwargs` to define any custom field-level validator, since it's more verbose and increases readability.

Performing object-level validation

We have learned how to perform field-level validation; now, let's learn how to perform object-level validation. This type of validation is particularly helpful when we have certain validations that involve multiple fields:

```python
class BlogCustom9Serializer(serializers.ModelSerializer):
    def validate(self, attrs):
        if attrs['title'] == attrs['content']:
            raise serializers.ValidationError('Title and content
cannot have value')
        return attrs

    class Meta:
```

```
        model = models.Blog
        fields = '__all__'
```

DRF executes object-level validation only when all the field-level validation is successfully executed. By defining our custom logic in the `validate()` method for serializers, we can perform object-level validation. For example, let's add a validation logic that will stop users from giving the title and content of the blog the same value:

Defining custom object-level validators

We can define custom object-level validators that can be used across multiple serializers. To do this, we need to declare the `validators` attribute in the `Meta` class. Here, we will implement the same logic as shown in the previous example – we want to prevent the same value in both the `title` and `content` fields. `Custom_obj_validator` is a custom object validator where we implement the logic:

```
def custom_obj_validator(attrs):
    print('custom object validator')
    if attrs['title'] == attrs['content']:
        raise serializers.ValidationError('Title and content cannot
have the same')
    return attrs


class BlogCustom10Serializer(serializers.ModelSerializer):
    class Meta:
        model = models.Blog
        fields = '__all__'
        validators = [custom_obj_validator]
```

We have seen multiple ways to implement validators at both the field level and object level, but it is important to understand the order of evaluation so that we can implement validators efficiently; we will discuss this next.

The order of the evaluation of validators

Let's explore the order of evaluation of different validators that will help us to decide where to add custom logic; we have used both field-level and object-level validator examples for clarity:

```
def func_validator(attr): # 1- Evaluates first
    print('func val')
    if '*' in attr:
        raise serializers.ValidationError('Illegal char')
    return attr
```

```
class BlogCustom11Serializer(serializers.ModelSerializer):
    def validate_title(self, value): # 2-If func_validator succeeds
        print('validate_title method')
        if '_' in value:
            raise serializers.ValidationError('Illegal char')
        return value

    def validate(self, attrs): # 3- If all field validator succeeds
        print('main validate method')
        return attrs

    class Meta:
        model = models.Blog
        fields = '__all__'
        extra_kwargs = {
            'title': {
                'validators': [func_validator]
            }
        }
```

The validator defined in extra_kwargs will always execute first (i.e., func_validator), followed by the field-level validate method (i.e., validate_title). If all the field validators are successful, then the object validator, the validate() method, will be executed. For better understanding, try to run the aforementioned example with different input values – for example, provide an invalid title value, and then add the correct title value while passing an object-level invalidate value.

Remove default validators from the DRF Serializer class

DRF Serializers have default validators that always check the input value. At a certain time, we might have use cases where we want to disable all the default field validators generated by DRF. If we assign the validator attribute an empty string, then all the automatically generated validators will be disabled.

For example, in the following snippet any field-level validator automatically generated is disabled for BlogSerializer, but if we define a validation in the validator method, then the validator method would still be applicable and executed for validation:

```
class BlogCustom12Serializer(serializers.ModelSerializer):
    class Meta:
        model = models.Blog
        fields = '__all__'
        validators = []
```

In this section, we have learned different ways to use DRF serializers for data validation. Field- and object-level validators have their own pros and cons, and developers should use them as per the validation requirements. Let's now start learning the intricacies of DRF serializers and how we can master different use cases for DRF Serializers.

Mastering DRF Serializers

We have learned all the important concepts of DRF serializers, so in this section, we shall learn how we can use DRF serializers efficiently, as well as a few new concepts that can help us master them even further.

Using source

We learned earlier how we can use the `source` option to perform multiple types of operations, and moving forward, you should use this argument in the serializer field whenever possible. This is because it drastically reduces the need to write manual code for basic tasks such as renaming fields, making a flat response structure, or accessing data from related fields.

Embracing SerializerMethodField

`SerializerMethodField` gives us a lot of flexibility to perform data manipulation on the fly. We should use method serializers whenever we need to perform computation involving multiple fields of a single object, or to transform data.

Using validators

Always try to add `validators` so that we don't need to use custom code to perform data validation. We have discussed multiple ways to add validators, both field-level and object-level in this chapter, and we should try to evaluate the appropriate type of validators and implement them accordingly.

Using to_internal_value

The `to_internal_value` method has access to raw input fields before validation is run. If we want to perform any operation on data or perform manipulation on the whole input data, we can make use of this method. Let's check out a quick example implementation of how to use it:

```
class BlogCustom13Serializer(serializers.ModelSerializer):
    def to_internal_value(self, data):
        print('before validation', data)
        return super().to_internal_value(data)

    class Meta:
```

```
model = models.Blog
fields = '__all__'
```

In this example, we simply add a print statement before calling the original implementation of the `to_internal_value` method that `ModelSerializer` provides out of the box. The `data` variable contains the raw input value, which can be manipulated as per our needs.

Using to_representation

The `to_representation` method takes a model object instance and converts it into a Python primitive datatype. We can override this method to perform data manipulation before the serialized data that the serializer instance returns. For example, in the `BlogCustom14Serializer`, we can convert the `title` field of the blog to uppercase text:

```
class BlogCustom14Serializer(serializers.ModelSerializer):
    def to_representation(self, instance):
        resp = super().to_representation(instance)
        resp['title'] = resp['title'].upper()
        return resp

    class Meta:
        model = models.Blog
        fields = '__all__'
```

In this example, we have made the title field uppercase. Similarly, we can perform any kind of operation on any of the fields before we return the response.

Use a context argument to pass information

In advanced use cases, we might need to pass the request object or any other additional information to the Serializer. A DRF serializer provides a `context` argument for this purpose. Any key-value pair passed through the `context` argument during the serializer initializations is available throughout all the serializer methods. This feature is handy when we want to pass any request object data to the serializer for custom computation.

In the aforementioned example, we pass a simple key-value pair to the `context` argument while initializing `BlogSerializer`:

```
class BlogCustom15Serializer(serializers.ModelSerializer):
    def to_internal_value(self, data):
        print('Printing context -', self.context)
        return super().to_internal_value(data)

    class Meta:
```

```
            model = models.Blog
            fields = '__all__'
>>> input_data = BlogSerializer(
    data={'title':'abc'}, context={'request': 'some value'}
)
>>> input_data.is_valid()
# Printing context - {'request': 'some value'}
```

The same key-value pair is accessible via `context` in the `to_internal_value` method. `self.context` is an object attribute; we can access the `self.context` value in any other method of the Serializer.

Customizing fields

DRF also provides a clean interface to create custom fields by subclassing the `Field` class or inheriting existing fields. The official documentation explains it quite well:

https://www.django-rest-framework.org/api-guide/fields/#custom-fields.

Passing a custom QuerySet to PrimaryKeyField

While creating a new related field, we might have a use case of only allowing `ForeignKey` for the fields that were created by the requested user in the past. For example, when we add a tag to a new blog, we want to make sure that only tags created by the author can be added to the blog. This is a very common use case that developers solve by writing custom code to validate the input data outside of Serializers.

However, instead, we can perform this operation by using `context` and the custom field concept discussed earlier. In the following example, we create a new `CustomPKRelatedField`, which overrides the default `get_queryset` behavior:

```
class CustomPKRelatedField(serializers.PrimaryKeyRelatedField):
    def get_queryset(self):
        req = self.context.get('request', None) #context value
        queryset = super().get_queryset() #retrieve default filter
        if not req:
            return None
        return queryset.filter(user=req.user) #additional filter


class BlogCustom16Serializer(serializers.ModelSerializer):
    tags = CustomPKRelatedField(queryset=models.Tags.objects.all())

    class Meta:
```

```
        model = models.Blog
        fields = '__all__'
```

Now, `CustomPKRelatedField` will automatically filter any tag created by the requested user and throw an error for any input request that has a tag outside the filtered entries.

Building DynamicFieldsSerializer

We have learned how to create different serializer fields and manipulate data using serializer concepts; however, in all our examples, the fields are always predefined. Perhaps there are use cases where we would want to add serializer fields dynamically; DRF does support this type of customization by overriding the regular behavior of how fields are defined in the Serializer class.

> **Read more**
>
> DRF also provides us with a way to create dynamic serializers that will generate fields on the fly. The official documentation has examples of how developers can use this feature better: `https://www.django-rest-framework.org/api-guide/serializers/#dynamically-modifying-fields`.

Avoiding the N+1 query problem

The N+1 query problem is a performance issue when the Serializer + ORM executes N additional SQL statements to fetch the same data that could have been retrieved when executing the primary SQL query. The N+1 query problem is a huge performance bottleneck that slows down a lot of applications in production, and it is quite common while using list Serializers that access related fields.

We can avoid this problem by using the `select_related` and `prefetch_related` querysets. It is advisable to solve this problem on a case-by-case basis and pass the appropriate Django ORM QuerySet to the serializer.

Now that we have mastered Serializers and the different configurations they provide, we shall focus on integrating DRF Serializers with DRF views.

Using Serializers with DRF views

In real-world application development, DRF Serializers work closely with DRF views to interact with the client application. Let's learn how we can integrate DRF Serializers to DRF Views by taking blog API examples. In the following example code, we integrate `BlogSerializer` to `APIView` for both read and write operations for the blogging model:

```
class BlogGetCreateView(views.APIView):
    def get(self, request):
        blogs_obj_list = Blog.objects.all()
```

```
        blogs = BlogSerializer(blogs_obj_list, many=True)
        return Response(blogs.data)

    def post(self, request):
        input_data = request.data
        b_obj = BlogSerializer(data=input_data)
        if b_obj.is_valid():
            b_obj.save()
            return Response(b_obj.data, status= HTTP_201_CREATED)
        return Response(b_obj.errors, status= HTTP_400_BAD_REQUEST)
```

In the preceding example, we list all the blogs when there is a get HTTP request, and in the post HTTP request, we create a new blog entry. Similarly, we can implement the business logic in the APIView class for other HTTP methods.

> **Important note**
>
> We shall directly focus on implementing advanced use cases. For basic integration, check out the DRF tutorial: https://www.django-rest-framework.org/tutorial/3-class-based-views/.

Working with generic views

Whenever we create simple CRUD operation APIs, it is always favorable to use generic views. Let's see how we can integrate Serializers into GenericAPIView. GenericAPIView has a serializer_class configuration via which we can link the associated Serializer.

For any write operation, DRF automatically takes care of passing the input field to the serializer and performing a save operation on the database. As we have learned earlier, we need to provide a QuerySet for serializers for any kind of read operation. In GenericAPIView, we can link the query using the queryset attribute configuration or by customizing the get_queryset() method. However, it is always preferable to use the get_queryset method to pass the queryset to the serializer.

Here's an example – let's create an API to create a new blog and list all the blogs that have an ID greater than 10:

```
class BlogGetUpdateView(generics.ListCreateAPIView):
    serializer_class = BlogSerializer

    def get_queryset(self):
        blogs_queryset = Blog.objects.filter(id__gt=1)
        return blogs_queryset
```

`BlogGetUpdateView` uses generic views to abstract out all the implementation details to get all the blogs lists and create a new blog API. We use `get_queryset` to pass the `queryset` value to our `BlogSerializer` serializer.

Read more

While implementing basic CRUD APIs, generic views are super helpful and faster to implement. DRF provides mixins for the custom creation of generic view behavior. The official documentation mentions all the available mixins that DRF provides out of the box:

`https://www.django-rest-framework.org/api-guide/generic-views/#mixins.`

Filtering with SearchFilter and OrderingFilter

We learned how we can customize a queryset for serializers using the `get_queryset` method; this method interface provides low-level control of what data we want to pass through the serializer. However, if we have a few generic use cases, such as ordering data as per certain fields or performing basic search operations, then we should use the `filter_backends` attribute. DRF provides two basic filtering backends, `SearchFilter` and `OrderingFilter`, to perform basic operations. For example, let us use `OrderingFilter` in `BlogGetUpdateView` to order all the results by the blog title:

```
class BlogGetUpdateFilterView(generics.ListAPIView):
    serializer_class = BlogSerializer
    filter_backends = [filters.OrderingFilter]
    ordering_fields = ['title']
```

Similarly, we can use the `SearchFilter` backend. The official documentation mentions basic examples of how you can use different filters and also how to create custom filter backends.

Read more

DRF also supports creating custom filters and expanding existing filters. The official documentation has examples, with proper explanations of how developers can make the best use of this feature:

`https://www.django-rest-framework.org/api-guide/filtering/#api-guide.`

In this section, we learned how to integrate DRF serializers with DRF views. We also saw how DRF supports filters out of the box and how we can use them to improve our development.

Summary

In this chapter, we learned how to work with DRF Serializers. First, we explored the concept of a basic DRF Serializer, as well as how DRF uses a Renderer and a Parser to transform data from JSON to Python native datatypes (and vice-versa). Then, we learned how to use DRF `ModelSerializer` with Django models.

DRF Serializers provide data validation support out of the box. In this chapter, we learned different ways to integrate field-level and object-level validation using serializers. We also learned how we can integrate DRF Serializers to work with Django-related fields and use nested Serializers.

Finally, we learned how to integrate DRF Serializers with DRF views to get an end-to-end picture of how Serializers work in the overall application development.

In the next chapter, we shall learn how we can use Django admin and work with Django management commands.

4

Exploring Django Admin and Management Commands

In *Chapter 3*, we learned how to integrate DRF serializers with Views and how client applications interact with Django Views. Our users would use the client application to perform CRUD operations using REST APIs. But in real-world applications, we would need to create a few additional interfaces to let admin users perform certain CRUD operations. Django provides an admin interface out of the box to cater to this use case.

In this chapter, we'll learn how to work with Django Admin, which provides us with a lot of flexibility to configure the admin interface so that it supports different business use cases. We'll learn how to explore those configurations, along with management commands to interact with the framework better (for example, to perform database migrations or start the development server). Django also provides an interface to extend these commands so that we can create custom management commands for custom use cases. We'll also take a look at this.

We'll cover the following topics in this chapter:

- Exploring Django Admin
- Customizing Django Admin
- Optimizing Django Admin for production
- Creating custom management commands

Technical requirements

In this chapter, we'll learn how to use the Django Admin panel. Django Admin works closely with Django models, so you're expected to have a good understanding of Django models.

The code snippets that will be used in this chapter can be found in this book's GitHub repository at https://github.com/PacktPublishing/Django-in-Production/tree/main/Chapter04.

Exploring Django Admin

For any application, we need some kind of admin UI to perform administrative tasks. Django Admin provides this powerful functionality out of the box. But before we can look at the interface, like any other admin interface, we need user login credentials to access Django Admin.

Creating a superuser in Django

Django provides us with the built-in createsuperuser management command to create new admin users. Once we run this command in our terminal, we shall see a guided wizard that will help us create a Django superuser:

```
python manage.py createsuperuser
```

By following the instructions and entering the data provided, we will get a Django superuser.

> **Django authentication and authorization**
>
> Django provides authentication and authorization support out of the box. When we are creating superuser in Django, we are going to use the default Django authentication system. We'll discuss the details of Django authentication and authorization in *Chapter 5*.

Once we have created a superuser, we can run our Django server to log into the admin interface. By default, Django Admin is configured in the /admin route, in the main urls.py file. Now, let's learn how to configure the Django Admin interface and all the features Django gives us to work with Django Admin.

Understanding the Django Admin interface

When we hit http://127.0.0.1:8000/admin in our browser, we will be redirected to the admin login page. The UI interface is autogenerated by Django and comes out of the box. Once we enter the credentials of our superuser admins we created earlier, we will be able to log into the admin interface, as shown in *Figure 4.1*:

Django administration   WELCOME, **ROOT**. VIEW SITE / CHANGE PASSWORD / LOG OUT

Authentication and Authorization administration

AUTHENTICATION AND AUTHORIZATION		
Groups	+ Add	⁄ Change
Users	+ Add	⁄ Change

Figure 4.1: Default Django Admin interface

At the moment, we have not configured anything extra. So, let's add a blog table interface to Django Admin. Every app has an `admin.py` file that helps us set up Django Admin for the given app. Then, Django Admin modules help us create user interfaces that can be used for adding/editing data to a database.

For example, once we add the following code to `blog/admin.py`, it registers the `BlogAdmin` class to the Django Admin UI. Here, we have linked `BlogAdmin` to the `Blog` model in our code example:

```
from django.contrib import admin
from blog import models


class BlogAdmin(admin.ModelAdmin):
    pass


admin.site.register(models.Blog, BlogAdmin)
```

Figure 4.2 shows the Django Admin UI with the new **BLOG** section added. As we add new admin classes and register them, they will start appearing in the UI:

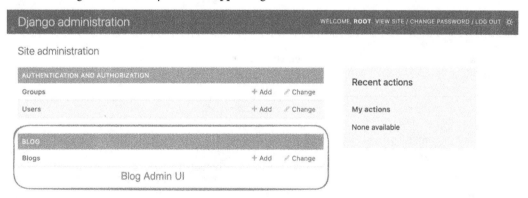

Figure 4.2: The BLOG section in the Django Admin UI

Now, let's configure Django Admin with more configurations that improve the user experience of the Admin pages. First, let's see what a list page for a blog model looks like in Django Admin. *Figure 4.3* shows the Django Admin list page for the blog model:

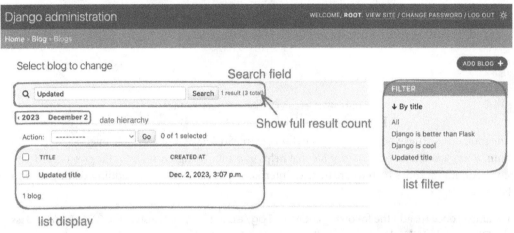

Figure 4.3: The Django Admin UI for the blog model

In *Figure 4.3*, we have highlighted the different types of configurations in Django Admin that are responsible for the UI. In the following snippet, we can see the attributes that are mapped to the relevant admin UI. The search_fields, show_full_result_count, and list_filter fields defined in the BlogAdmin class are responsible for the UI elements shown on the screen:

```
from django.contrib import admin
from blog import models


class BlogCustomAdmin(admin.ModelAdmin):
    search_fields = ['title']
    show_full_result_count = True
    list_filter = ['title']
    list_display = ['title', 'created_at']
    date_hierarchy = 'created_at'


admin.site.register(models.Blog, BlogCustomAdmin)
```

Similarly, we can configure the detailed UI for the blog admin page by defining more attributes for the BlogAdmin class.

> **Read more**
>
> The official Django documentation shows all the out-of-the-box options we can use to customize the Django Admin experience: https://docs.djangoproject.com/en/stable/ref/contrib/admin/.

So far, we have learned how to show the Django model fields in Django Admin. In the next section, we'll learn how to customize Django Admin with custom fields and other best practices.

Customizing Django Admin

In this section, we'll learn how to customize Django Admin to improve the user experience and solve different basic yet tricky requirements that come up from time to time.

Adding custom fields

Adding custom fields in Django Admin is one of the most common requirements when creating an interface. By doing so, we can add fields/information in Django Admin that are not present in the Django models. For example, to create a `word_count` field in Django Admin, we can use the following code:

```python
class BlogCustom2Admin(admin.ModelAdmin):
    list_display = ('title', 'word_count', 'id')

    def word_count(self, obj):
        return obj.content.split()

admin.site.register(models.Blog, BlogCustom2Admin)
```

In this example, we have added a custom field called `word_count` to `list_display` that is computed using other field values.

Using filter_horizontal

Django Admin enables us to use `ManyToManyField` in Django Admin. However, the default interface for `ManyToManyField` that's provided by Django Admin is not user-friendly.

As shown in *Figure 4.4*, it is difficult to identify the selected fields when we use the default admin interface:

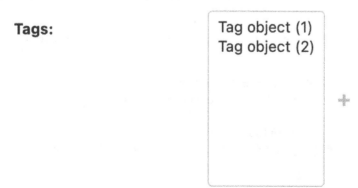

Figure 4.4: Default admin interface for ManyToManyField

To improve the admin user experience, we can pass the `tags` field to the `filter_horizontal` attribute in the `BlogAdmin` class. This converts the admin interface for `ManyToManyField` into a much better interface:

```
class BlogCustom3Admin(admin.ModelAdmin):
    filter_horizontal = ['tags']
admin.site.register(models.Blog, BlogCustom3Admin)
```

Using the `filter_horizontal` attribute, we can get a better interface, as shown in *Figure 4.5*. We can identify the fields that are selected and the ones that are available for selection:

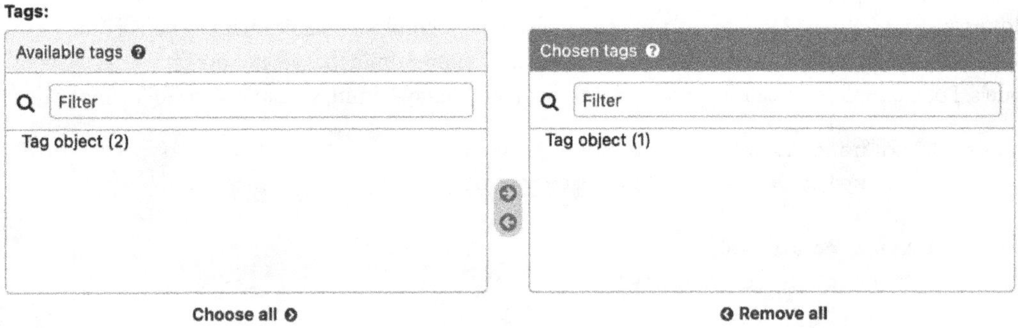

Figure 4.5: The admin interface when we use the filter_horizontal attribute

You can also use `filter_vertical` to improve the interface.

Using get_queryset

The data shown in the Django Admin interface is fetched via QuerySet. In *Chapter 2* and *Chapter 3*, we saw how we can gain a performance boost by using appropriate queries. While using Django Admin, we'll face the same N+1 select query problem that we discussed in *Chapter 3*. To solve this problem, we can override the default Django Admin QuerySet with a `select_related` or `prefetch_related` query.

In the following example, we are trying to fetch the full name of the user by accessing the `author.user` table data:

```
class BlogCustom4Admin(admin.ModelAdmin):
    list_display = ['title', 'created_at', 'author_full_name']

    def author_full_name(self, obj):
        return f'{obj.author.user.first_name} {obj.author.user.last_
name}'

    def get_queryset(self, request):
        default_qs = super().get_queryset(request)
```

```
        improved_qs = default_qs.select_related('author', 'author__
user')
        return improved_qs
admin.site.register(models.Blog, BlogCustom4Admin)
```

Since the author_full_name field has been added to list_display, Django Admin will make two queries per row. For example, if we have 10 blog entries, it would make 20 DB queries to fetch the author's name information. This takes a performance hit when we have hundreds of entries, so we are overriding the default get_queryset method to improve performance by using the select_related query.

Similarly, we can use different queries to override the default get_queryset method to gain performance. It is highly recommended to monitor the data access pattern for Django Admin as your application grows; these small performance bottlenecks can easily be solved.

Using third-party packages and themes

There are several third-party packages we can use to enhance the Django Admin interface. Though it is a debatable topic whether we should use third-party packages or not, I generally prefer using the default interface if Django Admin is used for basic CRUD operations for administrative purposes.

The following GitHub repository lists all the amazing Django Admin resources: https://github.com/originalankur/awesome-django-admin.

Using Django Admin logs

In any application, it is important to log admin actions performed by users. Django takes security very seriously and provides an admin action log interface out of the box.

By default, Django shows only the recently logged changes on the admin home page, as shown in *Figure 4.6*:

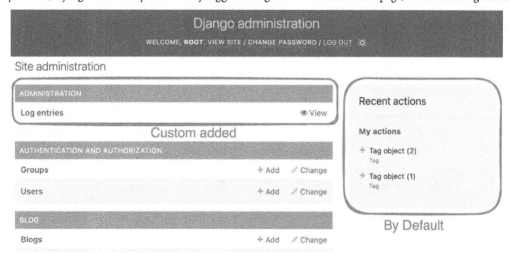

Figure 4.6: Showing the Django Admin action logs

Once we add the following code snippet to enable `LogEntryAdmin`, we are going to let users perform any read-only operation and block any write operation:

```
from django.contrib.admin.models import LogEntry

@admin.register(LogEntry)
class LogEntryAdmin(admin.ModelAdmin):
    def has_change_permission(self, request, obj=None):
        return False
    def has_delete_permission(self, request, obj=None):
        return False
    def has_add_permission(self, request):
        return False
```

We should ensure that `LogEntryAdmin` has restricted access and is read-only. Note that we have overridden the `has_<action>_permission` method so that users can only read the actions and not manipulate any of the past actions. Under the hood, Django stores all this log information in the `django_admin_log` table.

Now, let's start looking into how we can use Django Admin in production and learn all the best practices and tricks that help us use Django Admin better.

Optimizing Django Admin for production

While working with Django Admin in production, there are a few tips and tricks that you should follow to get the most out of Django. I have experienced a lot of challenges firsthand since developers bloat Django Admin with more and more administrative tasks. So let's look at a few good practices and optimizations that can help us use Django Admin to its full potential.

Renaming admin URLs

By default, Django Admin is configured to be accessible in the `/admin` path. However, Django Admin has access to all the data of your website; this is why it is important to move the path from `/admin` to some unguessable path, such as `/blog-cms`.

Different packages let us improve security. For example, **django-admin-honeypot** (`https://github.com/dmpayton/django-admin-honeypot`) is a package that creates fake admin pages and tricks attackers into accessing a dummy admin interface.

Using two-factor authentication (2FA) for admin users

Django Admin can be used by internal users to perform administrative tasks that need added privileges; since these users would have added permissions, safeguarding any security vulnerability should be the utmost priority. Adding 2FA for Django Admin users is very important in terms of security. We won't delve into implementing 2FA using Django here since these implementations are opinionated, but one of the recommended third-party packages that can help you set up 2FA in Django is **django-otp**, which you can find out more about here: `https://github.com/django-otp/django-otp`.

Using Custom Admin Paginator

The Django Admin interface has pagination implemented. As the data for a table grows, the performance of the default pagination will degrade. This is primarily due to the `count` query on a large table taking a longer time due to a large amount of data being present in the table; the query scans through all the rows to finalize the final count of the entries. We should override the default paginator with a custom paginator that overrides the `count` method. For example, we have created `CustomPaginator`, which overrides the `count` behavior:

```python
from django.core import paginator
from django.utils.functional import cached_property

# Idea referred from
# https://hakibenita.com/optimizing-the-django-admin-paginator
class CustomPaginator(paginator.Paginator):
    @cached_property
    def count(self):
        return 9999999


class BlogCustom5Admin(admin.ModelAdmin):
    paginator = CustomPaginator
admin.site.register(models.Blog, BlogCustom5Admin)
```

Now, Django Admin will not make a database call to fetch the total row count for a table and, by default, return a huge number (`9999999`), as defined in the code. We should only use this for large tables since we are returning a constant count number, which can be confusing to the end user.

Disabling ForeignKey drop-down options

`ForeignKey` fields in Django models will automatically add drop-down options in the Django model admin interface. This causes a performance bottleneck if we have a large table since Django will try to load all the foreign key objects in memory and show them in the drop-down option. Django understands the performance shortcomings of doing this, due to which it provides us with the `raw_id_fields` interface, which tells Django to replace the drop-down options with a free text field:

```python
class BlogCustom6Admin(admin.ModelAdmin):
    raw_id_fields = ('author',)
admin.site.register(models.Blog, BlogCustom6Admin)
```

Using list_select_related

When we use `list_display` to access the related fields, we get into the N+1 select query problem. Previously, we saw how we can override the default `get_queryset` method to perform optimization. But Django Admin provides the `list_select_related` attribute to tell Django to use a `select_related` query under the hood. As discussed in *Chapter 3*, using `select_related` to fetch information solves the issue of the N+1 select query.

For example, here, we are passing `author` and `author_user` as list values to `list_select_related`, which automatically performs select-related queries:

```
class BlogCustom7Admin(admin.ModelAdmin):
    list_display = ['title', 'created_at', 'author_full_name']
    list_select_related = ['author', 'author__name']

    def author_full_name(self, obj):
        return obj.author.name
admin.site.register(models.Blog, BlogCustom7Admin)
```

Please note that `list_select_related` doesn't work if we are trying to perform a reverse relationship. Hence, we need to override the `get_queryset` method, as we learned earlier.

Overriding get_queryset for performance

We should use `get_queryset` to optimize the queries used by Django. This is particularly helpful when we have reverse foreign key relationship lookup. We have discussed how we can override the `get_queryset` method in the *Using get_queryset* section.

Adding django-json-widget

Since Postgres and other databases support JSON fields, you will likely use JSON fields in Django. A lot of these fields will be manually accessed via Django Admin. The default UI that Django Admin provides for the JSON field is a `textarea` UI, but this is not user-friendly. One of the most useful third-party packages to solve this problem is `django-json-widget` (https://github.com/jmrivas86/django-json-widget). This package helps in getting a better interface for JSON fields.

Figure 4.7 shows the regular Django JSON editor in the admin panel:

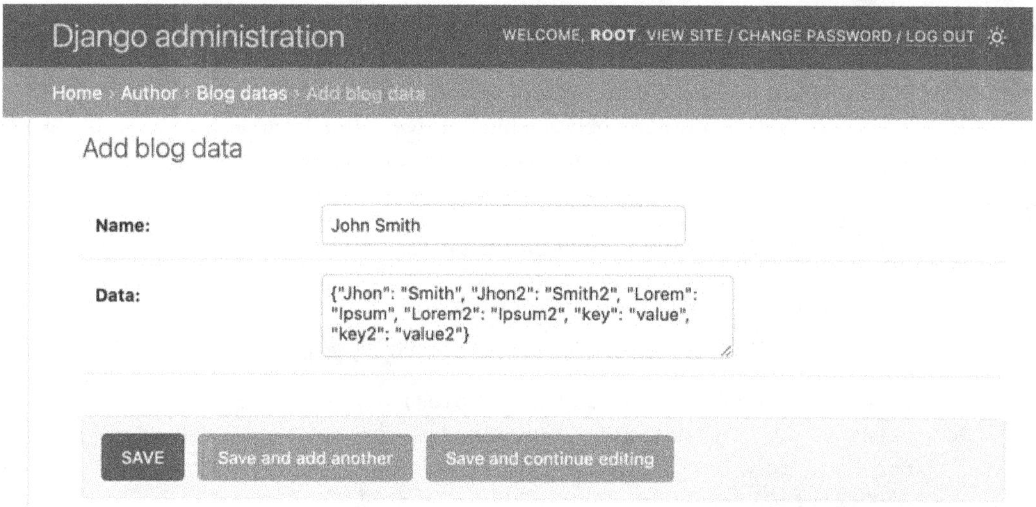

Figure 4.7: Showing the regular Django JSON editor

Now, if we use `django-json-widget`, we can use a much better JSON editor; its structure is shown in *Figure 4.8*:

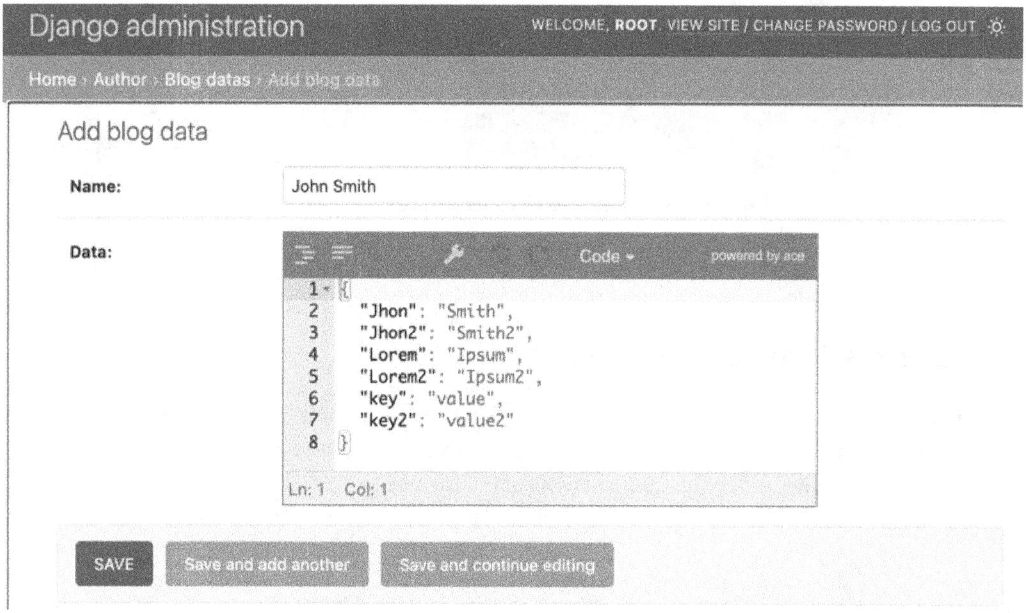

Figure 4.8: Showing the django-json-widget editor

Using custom actions

Often, a set of repetitive actions will be performed by an admin user using the admin interface. This will mean they will have to manually jump between multiple pages (for example, if we have to make a blog live, we would have to open each page and update the status flag for each blog). These kinds of operations can easily be performed using custom actions.

For example, let's create a custom action to print the titles of the selected blogs:

```
class BlogCustom8Admin(admin.ModelAdmin):
    actions = ('print_blogs_titles',)

    @admin.action(description='Prints title')
    def print_blogs_titles(self, request, queryset):
        for data in queryset.all():
            print(data.title)
admin.site.register(models.Blog, BlogCustom8Admin)
```

This code snippet gives us an additional drop-down option in Django Admin called **Prints title**, as shown in *Figure 4.9*. The new action option of **Prints title** will print all the blog titles in our terminal where the Django server is running:

Figure 4.9: Showing the newly added Action option called Prints title in Django Admin page.

Using permissions for Django Admin

The Django Admin interface helps us perform CRUD operations seamlessly using a UI. This also means that if someone has access to the Django Admin panel, they will have unrestricted access to user data. Django provides groups and permissions out of the box to help us with different permission levels in Django Admin. We'll how to efficiently use Django permissions in *Chapter 5*.

> **Avoid Django Admin for features**
>
> As projects mature, the feature requirements also evolve. Django Admin is most used for creating UIs for CRUD APIs for admin-related tasks. However, it should never be used as a client-facing feature. If you find yourself building features on top of Django Admin, then that is the right time to take a step back and build the UI in some other framework or platform.
>
> Django Admin can quickly become a permission nightmare if it's not managed properly. It is always recommended to use Django Admin when you are early in your product development stage. As your product/company matures, you should start thinking of migrating users away from Django Admin.

With that, we have learned how to use Django Admin in production and how we can configure Django Admin for better and more scalable usability. Now, let's learn how to create custom management commands in Django, which is used extensively.

Creating custom management commands

Django management commands help us interact with the Django framework seamlessly. A lot of tedious tasks such as migration, getting build files, and more become streamlined if we use Django management commands. Django provides a lot of management commands out of the box and can be found in the documentation: `https://docs.djangoproject.com/en/stable/ref/django-admin/`.

In this section, we are going to learn how to create custom management commands in Django. In Django, each application can register its custom actions with `manage.py`. To create a new management command, we need to create a simple file structure inside the Django app. In the given Django app, we need to create the `management/commands` directory structure, as shown in *Figure 4.10*. Any file that's created in the `commands` folder is registered as a Django management command, except if the name of the file begins with an underscore (`_`).

In the example shown in *Figure 4.10*, we have `publishedblogs` and `totalblogs` as two management commands, while `_private` would not be considered a management command:

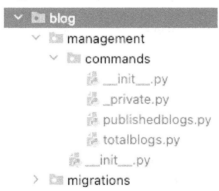

Figure 4.10: The folder structure for custom management commands

Please note that Django management commands will only be registered if the given app is added to the INSTALLED_APPS settings. Once we have created the management command files, we need to define a class called Command that is extended from BaseCommand in our custom management command files. For example, our total_blogs.py file would contain the following code:

```
from django.core.management.base import BaseCommand
from blog.models import Blog


class Command(BaseCommand):
    help = 'Returns total number of blogs'

    def handle(self, *args, **options):
        total_blogs = Blog.objects.count()
        self.stdout.write(self.style.SUCCESS(f'Total blogs: "{total_
blogs}"'))
```

The handle method computes the total number of blogs in the blog table and prints this out in the terminal using self.stdout.write(). We can execute our total_blogs custom command in the same manner, by using python manage.py total_blogs.

Sometimes, we will need to pass arguments to our custom command. We can do this by adding a new method called add_arguments to our Command class. For example, here, we are adding the add_arguments method, which configures two arguments that can be passed to our custom_command_info command. The first argument, author_ids, is a required positional argument, while custom is an optional argument:

```
class Command(BaseCommand):
    help = 'Returns total number of blogs'

    def add_arguments(self, parser):
        # Positional arguments
        parser.add_argument('custom_inputs', nargs='+', type=int)

        # Named (optional) arguments
        parser.add_argument(
            '--custom',
            type=int,
            help='custom optional param',
        )

    def handle(self, *args, **options):
        print(f'Custom inputs - {options["custom_inputs"]}')
        print(f'Custom - {options["custom"]}')
```

```
        total_blogs = Blog.objects.count()
        self.stdout.write(self.style.SUCCESS(f'Total blogs: "{total_
blogs}"'))
```

If we run the `custom_command_info` command, then we'll see the following output in our terminal:

```
> python manage.py custom_command_info 12 45 --custom 99
Custom inputs - [12, 45]
Custom - 99
Total blogs: "3"
```

Django management commands are fully customizable and we can even override the default management commands with our logic. Check out the official documentation for more detailed information on how to customize Django management commands: `https://docs.djangoproject.com/en/stable/howto/custom-management-commands/`.

Summary

In this chapter, we learned how to use Django Admin in our project. Django Admin is both customizable and scalable if we use the appropriate configuration, so we explored best practices to utilize its functions to their full capacity. We also learned when we should avoid using Django Admin and the performance issues developers face. Then, we learned about another powerful feature of working with Django, which is creating custom management commands.

Something we briefly learned about is how to create superusers in Django, but we didn't explore much about how authentication and authorization work. In the next chapter, we'll explore how we can use the Django authentication system better and how to expand it so that it supports different types of authentications, such as social auth, token-based auth, and more.

5

Mastering Django Authentication and Authorization

In *Chapter 4*, we learned how to use Django Admin to autogenerate a **user interface** (**UI**) for basic CRUD operations. While working with Django Admin, we used the default Django authentication and authorization systems. In this chapter, we will deep dive into the Django authentication system and learn how to use Django groups and permissions to implement authorization for resources. Django by default provides cookie-based session authentication, which works perfectly for browser-based applications, but when it comes to modern web apps, it is preferred to use token-based authentication. **Django REST framework** (**DRF**) provides token-based authentication out of the box, and we will learn how to integrate DRF token-based authorization into our project along with social login.

In this chapter, we will cover the following main topics:

- Learning the basics of Django authentication
- Customizing the `User` model
- Using a `OneToOneField` relationship with the `User` model
- Using Django permissions and groups
- Using DRF token-based authentication
- Learning about third-party token-based authentication packages
- Integrating social login into Django and DRF

Technical requirements

You should be familiar with the basic concepts of authentication and why authentication is required in web applications. We will be covering how to implement **role-based access control** (**RBAC**) in Django, so you are expected to have some basic understanding of the concepts of RBAC. You should also be familiar with the core concepts of **OAuth 2.0** that Google and other social login providers use to provide authentication services to different apps.

Here is the GitHub repository that has all the code and instructions for this chapter: `https://github.com/PacktPublishing/Django-in-Production/tree/main/Chapter05`

Learning the basics of Django authentication

The first concern any user has while using an application in today's world is, *Is my data secure with the application?* This is also the most important thing a developer should keep in mind – *always develop a secure application.* So, what does *secure* mean? Security can come into multiple levels, such as authentication, authorization, server security (that is, the server where the application is running is secure), and so on. In this chapter, we will focus on the most fundamental part of security, **application authentication**; that is, via a **login**. Django comes with a built-in authentication framework.

The batteries-included approach of Django provides an authentication system out of the box that gives us a basic *user model* that is plugged with session- (cookie)-based authentication. When we create a new Django project, by default, Django includes `django.contrib.auth` and `django.contrib.sessions` in the `INSTALLED_APPS` section of `settings.py`, which gives us the default `User` model and session-based authentication. In *Chapter 4*, we used the admin panel, which uses the same session-based authentication to provide a user-login workflow.

> **Read more**
>
> Session-based authentication is a vast topic in itself and beyond the scope of this book. If you are not familiar with the session-based authentication concept, please read the official documentation at `https://docs.djangoproject.com/en/stable/topics/http/sessions/`.

Django provides a standard `User` model out of the box, which comprises the following fields:

- `username`: Required field with a maximum 150 characters limit. It is used for login purposes. Please remember – this is not the email address, so one can use this field as a unique identifier for login purposes. I have seen organizations use this field as email, phone number, or custom usernames, so feel free to utilize this field as any field.

- `first_name`: Optional field with a maximum 150 characters limit.

- `last_name`: Optional field with a maximum 150 characters limit.

- `email`: Optional field that can be used to capture the email address of the user.

- `password`: Required field that saves the hashed value of the set password. Django doesn't store the raw password in the database.

- `is_staff`: A Boolean field that is used to check if the user can access the admin site or not.

- `is_active`: A Boolean field that is used to identify if the user account is enabled or not. If the `is_active` field is set to `False` then the default `ModelBackend` authentication provided by Django would not let the user log in. *Please note that if you are creating any custom login system, then it would not be respected unless you have implemented a custom logic to check the flag.*

- `is_superuser`: A Boolean field that should be used cautiously. If this flag is set for any user, then they have full access to all of the Django admin sections.

- `last_login`: Datetime field that captures when the user was last logged in. *Please note this field is updated only when we use the default login mechanism by* `ModelBackend`.

- `date_joined`: Datetime field that captures the datetime when the user account was created.

- `groups`: Many-to-many relationship with the `Group` table. (We will discuss the *groups* concept in the *Using Django permissions and groups* section.) This is particularly helpful when we use Django authorization to provide access control to users. Django provides RBAC support using this field in the backend.

- `user_permissions`: Many-to-many relationship with the `Permission` table. Django uses permissions in combination with groups to provide RBAC on the admin portal or at the API-level permission using the `has_perm` method. (We will discuss the *permissions* concept in the *Using Django permissions and groups* section.)

All the aforementioned fields are standard and would suffice for any basic project. Many organizations or projects might have a custom requirement that would need additional fields associated with a given user, and that is when we can customize the `User` model. In the following section, we will learn about the different approaches one can take to customize the `User` model in Django.

Customizing the User model

There are primarily two approaches via which we can customize the `User` model. Both of them have their equal advantages, so let us learn about them one by one:

- **Extending the User model by a one-to-one relation**: Whenever there is a requirement to add additional fields to the user, this is one of the simplest and best solutions. In this approach, we would create a new model, say `UserProfile`, and then create a one-to-one relationship with the `User` model. Any new custom field would be added to the `UserProfile` model. I would recommend using this approach most of the time since this is one of the simplest and cleanest solutions that doesn't need much effort from the developer's end and serves our requirements.

- **Expand using AbstractUser and AbstractBaseUser model**: This is another approach to expand the `User` model. In this approach, we would have to create our custom `User` model and inherit from the `AbstractUser` class or `AbstractBaseUser` class. The key differences between these models are:

  - `AbstractUser`: Use the `AbstractUser` model only if you want to update the username field in the `User` table.

  - `AbstractBaseUser`: Use the `AbstractBaseUser` model if you want to create all fields in the `User` table from scratch.

  It is recommended to use the custom model approach only when you have a good understanding of Django and are familiar with the login and authentication workflow.

Now, let us learn to create a custom `User` model by using the `AbstractUser` or `AbstractBaseUser` model in our Django project:

1. Create a new Django app, `custom_user`.

2. Then, define a `CustomUser` model with the appropriate fields, such as `phone_no` and `city`, which inherit the `AbstractUser` class.

3. Next, set the `username` field to `None`. We also need to define a custom model manager from `BaseUserManager`:

```python
from django.contrib.auth.base_user import BaseUserManager
from django.contrib.auth.models import AbstractUser
from django.db import models
class CustomUserManager(BaseUserManager):
    """
    Used for updating default create user behaviour
    """

    def create_user(self, phone_no, password, **kwargs):
        # implement create user logic
        user = self.model(phone_no=phone_no, **kwargs)
        user.set_password(password)
        user.save()
        return user
    def create_superuser(self, phone_no, password, **kwargs):
        # creates superuser.
```

```
        self.create_user(phone_no, password)

class CustomUser(AbstractUser):
    username = None
    phone_no = models.CharField(unique=True, max_length=20)
    city = models.CharField(max_length=40)
    USERNAME_FIELD = "phone_no"
    objects = CustomUserManager()
```

The preceding code snippet is defined in the custom_user/models.py file. In the code snippet, we have created a CustomUser model and a CustomUserManager model manager, and we have updated the username logic to phone_no. This means, going forward, we can utilize the phone_no field to log in to our Django application, along with the password.

4. Once we have our model created, we need to tell Django to utilize the new model for any authentication purpose. We need to go to our settings.py file and then define a new variable:

```
AUTH_USER_MODEL = "custom_user.CustomUser"
```

Please note that "custom_user" is our custom app name where the CustomUser model is defined.

5. Now, run the Django migration command to create our new User model.

6. We also need to update the Django Admin forms so that our Django Admin tool is working. We will not get too deep into the details of form, since this is out of the scope of this book.

> **Read more**
>
> Django's official documentation provides a well-defined example; please go through this at https://docs.djangoproject.com/en/stable/topics/auth/customizing/#a-full-example.

We've learned how Django provides a User model out of the box and how we can customize the User model to fit our use case.

Though we have learned how to use a custom User model in Django, it is recommended to create a separate profile model to capture all the additional user information and *use a one-to-one relationship* with the default User model so that Django can take care of all the authentication and other security features. Using this approach, developers can focus on product feature development without getting too much involved in framework-level changes for authentication.

Using a OneToOneField relationship with the User model

One of the simplest ways to store custom user information in Django is by creating a new model and creating a `OneToOneField` relation with the default `User` model. Let us take a simple example to implement this approach.

We need to store the phone number and city for each user registering on our platform, but Django's default `User` model does not have these fields present. We should create a new model to store this information for each user.

First, create a new custom Django app, `custom_user`. Then, add a new `UserProfile` model in the `custom_user/models.py` file:

```
from django.db import models

class UserProfile(models.Model):
    user = models.OneToOneField('auth.User', related_name='user_
profile', on_delete=models.CASCADE)
    phone_no = models.CharField(unique=True, max_length=20)
    city = models.CharField(max_length=40)

    def __str__(self):
        return self.user.username
```

Here, we have created a `user` field that has a one-to-one relationship with Django's default `User` model. The phone number and city information of each user is stored in the `phone_no` and `city` fields.

The biggest advantage of using this approach of creating a custom model such as `UserProfile` is that we can have any field and logic that would not clash with Django's default `User` model behavior.

We saw how the Django `User` model has a relationship with *groups and permissions*. In the next section, we will deep dive into the details of the `Group` and `Permission` tables that can help us provide RBAC support to end users.

Using Django permissions and groups

RBAC is a method of restricting access based on roles assigned to individual users. Django permissions and groups are some of the most thought-through and verbose RBAC systems I have come across in my career. One reason why I always choose Django for any tight-deadline project is primarily due to the authentication and authorization system it provides out of the box. In this section, we will get a high-level overview of how we can use Django groups and permissions in our project.

Using permissions and groups in Django Admin

In *Chapter 4*, while exploring Django Admin, we used a Django **superuser** to navigate through the admin panel; hence, there was no permission needed. But as our project moves to production, we want to give restricted access to each user depending upon their use case. For example, a support agent would need *view-only* access to all payment models and should not have access to any other database. These kinds of RBACs can easily be achieved by Django Admin.

When we log in to the Django admin panel as a superuser, we will be able to see the authentication and authorization UI, as shown in *Figure 5.1*. The **Groups** and **Users** sections help us to have RBAC for our Django application:

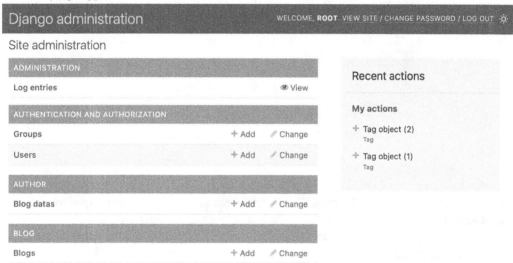

Figure 5.1: Django admin panel showing AUTHENTICATION and AUTHORIZATION

By default, Django creates permission objects for each model object. These permission objects are used in Django Admin to check whether a particular user using Django Admin has authorization to perform the requested model operation. If we open an auth `User` object in the Django Admin UI, we would be able to see all the groups and user permissions attached to the particular user. For example, as shown in *Figure 5.2*, we can see all user permissions available to be attached to the user. Django creates permission objects for all models available in the project, with different operation types such as `add`, `change`, `delete`, and `view`. We can select a particular model and then attach the required permission to the user object:

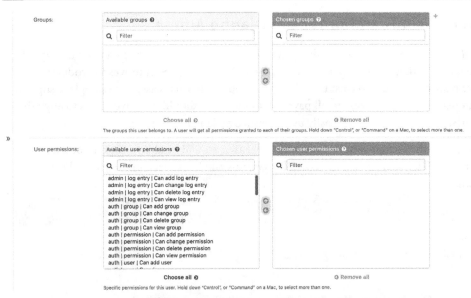

Figure 5.2: User object view in Django Admin showing all the groups
and user permissions attached to the given user

Though we can always attach individual permission to users, it is always advisable to use groups to give permissions to any user so that all permissions are managed centrally via groups. Django **permissions** help us create granular control of model operations, and Django **groups** are collections of permissions that help us create roles. For example, a support agent might need access to only view the author model and blog model, so we would create a Django group named **Support Agent** and add author and blog model view-only permission, as shown in *Figure 5.3*:

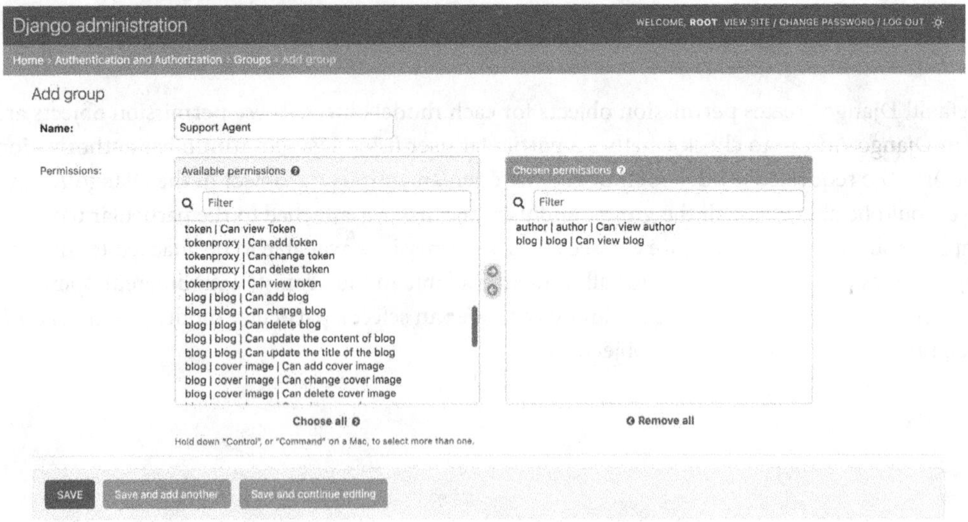

Figure 5.3: Creating a new Django Support Agent group

Once we have the Django **Support Agent** group created, we attach the group to the user, as shown in *Figure 5.4*. This will allow the user to have view-only access to the `author` and `blog` table via Django Admin:

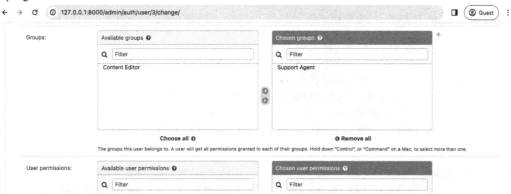

Figure 5.4: Attaching the Support Agent group to a Django user

Using Django groups helps us control permissions of the Django Admin interface from a central place for each type of user. For example, if we want to remove access to viewing author information for all support agents, we will go to the **Support Agent** group and remove the author view permission attached; this would automatically remove access to the author model from all support agents who were given access by the Django **Support Agent** group.

Django provides basic CRUD operation permission objects out of the box, but our business use case might need custom permissions; for example, only certain users can update the title of a blog while others can only edit the content of the blog. For such use cases, Django allows us to create custom permissions.

Creating custom permissions

Django permissions are at the model level. To create a custom permission, we have to use the permissions model `Meta` attribute. For example, we have to create permissions to allow certain users to only update the title of a blog and certain users to update the content of the blog:

```
class Blog(models.Model):
  ...
  class Meta:
    permissions = [
       ("update_title", "Can update the title of the blog"),
       ("update_content", "Can update the content of blog"),
    ]
```

Once we define the permissions in our model, we would need to run Django migration to create the permissions object.

Just creating custom permissions would not mean that our application would have all the custom permissions implemented. We need to write custom logic to enforce the authorization using Django permissions.

Let us create a simple Django view that would use the custom permissions and check the result:

```
def update_blog_title(request):
    blog_id = request.GET.get('id')
    blog = Blog.objects.get(id=blog_id)
    if request.user.has_perm("blog.update_title"):
        return HttpResponse('User has permission to update title')
    return HttpResponse('User does not have permission to update
title')
```

Now, if we tried to access the view on our browser, we would get a `User does not have permission to update title` message. But if we added the permission to a user and then tried to access the page, then we would get a `User has permission to update title` response.

Please note

If you are logged in to the Django Admin interface as a superuser, then you will always see `User has permission to update title`. This happens because, in Django, the superuser has access to all permissions implicitly. However, not every staff user has site-wide permission, so one should create a normal staff user and assign custom permissions to test the code.

Django custom permissions enable us to utilize the existing Django permissions framework to have custom RBAC support for our Django application.

Using Django permissions and groups for an API

Django groups and permissions can be used for an API by writing custom code. DRF also supports a custom `Permission` class that can use Django permissions and groups under the hood:

```
def check_permission(user, group_name):
  return user.groups.filter(name=group_name).exists()


@api_view(['POST'])
def blog_view(request)
  if not check_permission(request.user, 'can_view_blog')
    return Response(status=403)
  # perform operation
```

In the preceding code example, we are checking if the user making the request is eligible to comment or not. If the `can_view_blog` group is attached to the user, then they are eligible to comment; otherwise, they are not eligible to comment, and we will respond with a `403` error.

Similarly, we can write custom business logic in our code to support authorization in our DRF API. We can also write a custom DRF permission class to perform such operations.

> **Read more**
>
> For more details, check the official documentation on how to create a custom permission class in DRF: `https://www.django-rest-framework.org/api-guide/permissions/#custom-permissions`

Caveats of using permissions

Though Django groups and permissions are one of the most powerful RBAC systems I have come across, I have also seen a lot of projects becoming nightmares due to the overuse of the Django permissions framework. In this section, we will try to learn all the good practices that developers and teams can adapt to use Django groups and permissions to their advantage and in the right way:

- **Avoid attaching permissions directly to auth user objects**: When we start using Django permissions, it might be tempting to directly attach permissions to django auth users, but this temptation soon leads to a nightmare. As we start using more Django permissions, it becomes tricky to keep track of which user has access to which part of the Django application. To solve this problem, always use Django groups to attach permissions to users.

- **Avoid creating superusers**: We should avoid creating superusers for our project because they have unrestricted access to perform any operation using Django Admin.

- **Avoid using too many Django custom permissions**: Django permissions are attached to a model, and a lot of times, developers tend to unnecessarily attach custom Django permissions to models when they don't need any model relation. Try to use Django groups for this purpose.

  In the following code snippet, we are checking if a user is attached to a particular group for authorization:

  ```
  if user.groups.filter(name='custom_group').exists():
      # perform operation
  ```

  `custom_group` might not have any model permission attached to it, but a basic Django group can be used to check if the user belongs to a particular group to operate.

So far, we learned about the Django permissions framework and how we can use it to create an RBAC system in our Django application. Now, we will look into how DRF supports token-based authentication and how we can integrate it into our Django project.

Using DRF token-based authentication

Modern web applications are no longer limited to just browsers. In fact, *more than 55%* of internet traffic comes from mobile (as per this *Cloudflare* report: `https://radar.cloudflare.com/traffic?range=28d`). Django was first built more than 18 years ago when browsers were more prevalent, but in today's world, there are more mobile devices and applications than browsers, which require a different type of authentication. DRF already provides multiple types of authentications out of the box:

- Basic authentication
- Session-based authentication
- Token-based authentication
- Remote user authentication

Token-based authentication is the most popular and widely used form of authentication in today's world. In this section, we will learn how to enable and use token-based authentication in our Django project, the advantages of using token-based authentication, and another custom type of token-based authentication.

Integrating token-based authentication into DRF

DRF token-based authentication uses a simple token-based HTTP authentication schema. DRF creates a new model that saves the issued token for a given user and has a one-to-one relationship with the User table. Let us check step by step how DRF implements the token-based authentication workflow:

1. The user makes a login request that contains a username and password.
2. Django verifies the username and password from the `Auth User` table.
3. Once the user credentials are verified, DRF creates a new token for the given user and then sends the token to the end user as a response.

   All subsequent requests made by the client now have the issued token attached to the HTTP header.

4. When Django receives a request that has the auth token attached to the request, Django then validates the token against the saved token from the database and attaches the `Auth User` object to the `request.user` property. This hydrates the user details to the `request` object, and the token can be used to identify which user has made the request. If the token is mismatched, then DRF will send a `401` error to the client.

In this section, we learned on a high level how DRF token-based authentication works. Now, let us look into the integration of token-based authentication into Django using DRF.

Adding DRF token-based authentication to a Django project

In DRF, authentication is checked as the first step before the execution of the view. It is important to understand that, unlike Django session-based authentication, DRF authentication is *not implemented via middleware*. Instead, DRF checks for authentication when the associated view execution starts. DRF takes this approach primarily to make sure DRF is a plug-and-play system that can easily co-exist with other authentication mechanisms used by other views. For example, Django Admin Views uses the native Django session-based authentication.

Just as with Django, DRF also attaches the Auth User object to the request.user attribute after successful authentication of the request. If the request does not have any authentication attribute, then it attaches an Anonymous user to request.user. Let us now learn the steps to integrate a DRF token into our Django project:

1. The DRF package has an internal app that creates a Token table in our database. We need to add rest_framework.authtoken to the installed apps of our Django settings to attach the authtoken app to our project:

```
INSTALLED_APPS = [
    ...
    'rest_framework.authtoken'
]
```

2. Run the python manage.py migrate Django migration command to create the Token table in our database.

3. Add the default authentication class to the configuration settings.

   DRF evaluates the authentication before the execution of the view starts, so we can define the authentication class for each view. But this could cause security gaps if the developer forgets to add authentication. Hence, it is advisable to attach the default authentication class. If we need to change the authentication for a particular view, then we can always override the authentication in the *view* definition.

   Just attaching authentication to a view is not enough; we also need to tell DRF what the permission type for the given view is. By default, we should add IsAuthenticated to all our views so that all the views are always accessible by the authenticated users unless explicitly mentioned.

   We need to add the following configuration to our settings.py file to enable authentication and permission to all our views:

```
REST_FRAMEWORK = {
    'DEFAULT_AUTHENTICATION_CLASSES': [
        'rest_framework.authentication.TokenAuthentication',
    ],
    'DEFAULT_PERMISSION_CLASSES': [
```

```
                        'rest_framework.permissions.IsAuthenticated',
        ]
    }
```

Every request that would be mapped to the DRF view would now go through DRF token-based authentication, and we would check whether the user making the request is already logged in or not.

4. We need to create a `login` view that would let the users access the token for authentication. Please note that we need to explicitly add `permission_classes` as `AllowAny` for our `login` view, or else we will get a `401` error:

```python
from rest_framework.authtoken.models import Token
from rest_framework.permissions import AllowAny
from django.contrib.auth import authenticate

@api_view(['POST'])
@permission_classes([AllowAny])
def login(request)
    username = request.DATA['username']
    password = request.DATA['password']
    user = authenticate(username=username, password=password)
    if not user:
        return Response(status='401')
    token = Token.objects.get_or_create(user=user)
    return Response(data={"token": token.key})
```

5. Once we generate the token, we would need to attach the token to every subsequent request made to our Django server. Add the token to the `Authorization` header key; the key should be prefixed by the `Token` string literal, with whitespace separating the two strings:

```
Authorization: Token <token key>
```

Then, the `curl` request should look something like this:

```
curl -X GET http://127.0.0.1:8000/api/example/ -H
'Authorization: Token 9944b09199c62bcf9418ad846dd0e4bbdfc6ee4b'
```

We now have DRF token-based authentication configured for our project. With this, any request made to our DRF views would be always checked for token-based authentication and whether the request is authenticated or not. DRF also supports parallel authentication and permission classes; if we want to attach more authentication methods, then add them as a list. For example, if we want to also allow the default Django session-based authentication to work for our DRF views, then we can add the following:

```python
REST_FRAMEWORK = {
    'DEFAULT_AUTHENTICATION_CLASSES': [
        'rest_framework.authentication.TokenAuthentication',
```

```
                'rest_framework.authentication.SessionAuthentication',
        ]
}
```

Using multiple Django authentications can help us support multiple authentications for a Django project. DRF would try to authenticate the request using the first authentication class. If successful, then it would not check the other classes and attach the user to the `request.user` attribute, or else, it would keep trying the subsequent authentication classes.

> **Read more**
>
> The DRF permission class uses the Django groups and permissions framework under the hood to support model-level permission. We are not getting into the details of the Django permissions framework and how one can use it in a DRF project. You can go through the examples mentioned in the DRF documentation for more details: `https://www.django-rest-framework.org/api-guide/permissions/#api-reference`.

DRF also provides an interface to create custom authentication and permission classes as per our requirements. The official documentation of DRF mentions step-by-step processes to expand these classes. We will not get too deep into these classes, and I'll let you folks follow the official tutorial for this purpose.

> **Read more**
>
> For the custom authentication class, follow this link: `https://www.django-rest-framework.org/api-guide/authentication/#custom-authentication`.
>
> For the custom permission class, follow the tutorial mentioned in the official documentation at `https://www.django-rest-framework.org/api-guide/permissions/#custom-permissions`.

Now, let us look into a few limitations of DRF token-based authentication.

Understanding the limitations of token-based authentication of DRF

DRF token-based authentication is a fairly simple implementation. For any basic application, DRF works like a charm, but as you start building complex features with authentication, then one finds the pitfalls. Let us understand some shortcomings of DRF token-based authentication:

- **No support for multiple device login**: Since DRF creates a single token for each user, if a user logs into multiple devices, then we would be using the same token for each device. This is a security concern if the user has logged in to an untrusted device. The only solution left for the user is to delete the old token; this would log out the user from every device.

- **Unencrypted tokens are stored in a database**: Whenever a new token is created, DRF saves the token in raw format; hence if the database ever gets compromised, a hacker would be able to get access to all users.

- **No expiry**: DRF tracks the creation time of each token, but it does not provide support to expire the tokens. With modern applications, it is very important to provide expiry and refresh token features.

- **Database lookup for every request**: For every request, there would be a database query involved to fetch the information of the user against the token set in the header.

- **No way to send any data to the frontend via the token**: Modern-day applications use tokens to retrieve some basic information about the user without making an additional API call.

- **No support for social login**: Users want a smooth single-click sign-in process via social logins platforms such as Google, GitHub, and so on.

These are a few common disadvantages of using DRF token-based authentication. Though there are no simple solutions that can solve all these use cases, we can look into a few third-party packages that provide different approaches to solving these problems.

Learning about third-party token-based authentication packages

In this section, we will get a high-level overview of different packages available that can be plugged into Django and DRF to solve the challenges of DRF token-based authentication. We will not deep dive into the implementation and integration of the following package; rather, we will enable developers to know about the available open source packages out there that can be easily used by them to solve problems.

django-rest-knox

The `django-rest-knox` package provides multi-device login and session management support. It has a similar architecture to DRF token-based authentication, but it solves a couple of issues, such as saving tokens in an encrypted format and also providing expiry time for issued tokens.

> **Read more**
> For more details, check the official documentation at `https://jazzband.github.io/django-rest-knox/`.

djangorestframework-simplejwt

JSON Web Token (JWT) is a modern-age token-based authentication standard that is fairly popular among developers (read more at `https://jwt.io/introduction`). The advantage of JWT is that it doesn't need any database lookup for validation, and we can also add information to the issued JWT token itself that can be utilized by the client to get information about the user. The `djangorestframework-simplejwt` package helps to enable JWT authentication in the DRF project.

> **Read more**
>
> The official documentation has detailed instructions on how one can integrate JWT into a Django project: `https://django-rest-framework-simplejwt.readthedocs.io/en/latest/getting_started.html`

There are plenty of packages, such as `django-allauth`, `djoser`, `django-rest-knox`, `dj-rest-auth`, and so on available that provide a different set of authentication mechanisms, but before choosing any package, we should be aware of what features the package brings to the table and whether it makes sense to build the features ourselves. This is a trade-off that needs to be calculated because once we use any of the third-party packages for authentication, it will be very hard to reverse that decision.

One last important aspect of authentication is social login. Social login is the most common form of login in today's modern web applications. Let us learn how we can incorporate social login for our Django project.

Integrating social login into Django and DRF

Social login is crucial for any client-facing application. In today's modern world, users do not want to fill out lengthy forms to register, nor do they want to remember passwords for every site; these hurdles can be resolved using social login. There are multiple services, such as Google, Facebook, Twitter, and GitHub, that can be plugged into our website to enable social login. Each of these platforms has different implementation details, and it can be tricky to implement them. I recommend you use the `python-social-auth` third-party package to integrate social login into the Django application.

We will not get into the implementation details for `python-social-auth` (`https://python-social-auth.readthedocs.io/en/latest/configuration/django.html`) since there are multiple types of service providers available, such as Google, Facebook, GitHub, and so on. You are advised to go through the integration documentation for their respective platforms.

Summary

In this chapter, we have learned how Django provides authentication out of the box. Django also provides authorization support with the help of Django permissions and groups. RBAC is one of the key security features that is needed by modern applications, which is provided out of the box by Django.

We have also learned how to use token-based authentication using DRF and Django. Token-based authentication is useful for non-browser clients, such as Android, iOS, and IoT devices. Social login is a must-have feature in today's modern applications, and we have discussed how `python-social-auth` in Django can be used to integrate social login.

In the next chapter, we will learn more about how we can implement caching, logging, and throttling in Django applications and the added advantages they offer.

Part 2 –
Using the Advanced Concepts
of Django

In this part, we can expect to learn advanced concepts in Django such as caching, pagination, Django signals, and middleware. The Celery framework is one of the most popular frameworks used for asynchronous tasks, and we shall learn how to integrate Celery into a Django project in this part. We will also learn how to write test cases for our Django project, as well as what the best practices and conventions used across the software industry are while working with Django.

This part has the following chapters:

- *Chapter 6, Caching, Logging, and Throttling*
- *Chapter 7, Using Pagination, Django Signals, and Custom Middleware*
- *Chapter 8, Using Celery with Django*
- *Chapter 9, Writing Tests in Django*
- *Chapter 10, Exploring Conventions in Django*

6

Caching, Logging, and Throttling

In *Chapter 5*, we learned how we can authenticate every request and perform authorization to make sure the client has appropriate permission to access resources. These operations make additional DB calls for each request. For most use cases, these additional DB calls are insignificant. But as we build the system, we might come across use cases where we want faster response times for DB read operations; caching plays an important role when we try to achieve this. In this chapter, we will learn how to use caching in Django. Throttling is the process of limiting the number of requests made by the client in a certain period. Logging and throttling are advanced concepts that we use extensively in the industry.

In this chapter, we will learn:

- Caching with Django
- Logging with Django
- Throttling with Django

Technical requirements

We are expecting the readers to have a basic idea of what caching means and how it is beneficial to end users. We are expecting our readers to be familiar with the concepts related to and the use cases for implementing throttling.

Both caching and throttling need an additional in-memory database: **Redis**. Though we shall use abstracted Redis APIs for our implementation, it is recommended that the reader knows what Redis is and its basic concepts. Apart from this, we are going to learn about how one can perform logging operations in Django. The Django logging framework uses and extends Python's built-in logging module, so we are expecting our readers to know how the Python logging module works on a high level.

> **Learn more**
>
> To get a basic understanding of Redis, one can follow the official documentation of Redis or take one of the Redis courses available at `https://university.redis.com/`.
>
> Python official documentation has a fairly simple guide to learning more about the built-in logging module at `https://docs.python.org/3/library/logging.html`.

Caching with Django

Caching is considered one of the necessary evils of computer science. It helps to attain performance gain, but it also makes the system complex. One of the most important complexities in computer science is invalidating the cache. In this section, we will learn how caching is implemented in Django and also how we can improve the overall caching experience for developers using Django and DRF.

Django provides different types of caching strategies out of the box:

- **Local memory caching**: The data is cached in the application memory itself.
- **Database caching**: Django would store the cached results in the database itself.
- **Filesystem caching**: Cached entries are saved as files on the file system.
- **Memory database-based caching**: Django supports caching databases such as Memcache and Redis to be used in the project. We shall learn how to integrate Redis into our Django project in this chapter.
- **Custom caching**: Developers can also write a custom backend to create their own strategy to integrate with Django.

From Django 4 onwards, Django supports Redis as an in-memory database for caching out of the box. Let us first learn how to use the Redis database. Just like Postgres, we have to install Redis on our system to use it. However, we don't want to get into the hassle of setting up Redis on our system; hence, we shall use Redis cloud service in this chapter. Later, in *Chapter 11*, we will learn to use Redis with Docker.

> **Important note**
>
> We are not going to deep dive into Redis configuration and setup in this chapter. We shall primarily focus on integrating Redis into a Django project and how we can use caching in our project. As the project grows, developers would need to fine tune the cache configurations further to handle the scale and advanced use cases.

To use Redis here, visit `https://app.redislabs.com/#/login`, create a new account, and follow the wizard to create a new Redis database. *Figure 6.1* shows the Redis labs dashboard, which we shall use:

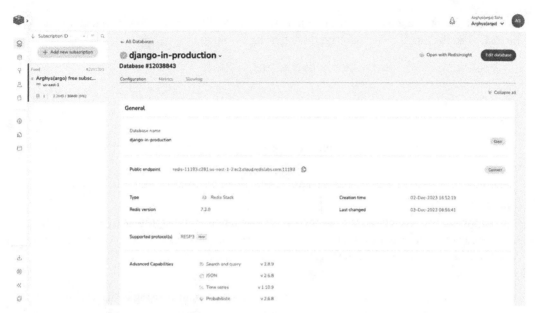

Figure 6.1: Redis dashboard from RedisLabs

Once we have created the Redis database, we need to identify the following information from the RedisLabs dashboard:

- **Public endpoint**: This has the information about the host URL and port
- **Redis user**: For the free plan, the user is always **default**
- **Redis password**: This is mentioned in the **Default user password** field

Now let us check how we can use the configuration and add Redis caching support to Django.

Add the following configuration to the `settings.py` files:

```
CACHES = {
    'default': {
        'BACKEND': 'django.core.cache.backends.redis.RedisCache',
        'LOCATION': 'redis://<user>:<password>@<public endpoint>',
    }
}
```

Apart from configuring in the `settings.py` file, we would have to install the `redis` package using `pip`. We can do so by `pip install redis`.

Once you add the configuration, we need to test whether our Redis configuration is valid and working. To do that, open the Django shell and run the following command:

```
>>> from django.core.cache import cache
>>> cache.set('hello', 'World', 600)
>>> cache.get('hello')
'World'
```

If we see the output `World`, then it is confirmed that our caching configuration in Django is working as expected.

> **Important note**
>
> Redis support was added to Django only from Django 4 and above. If you are using lower versions, you can consider upgrading the Django version or using `django-cacheops`, which we have discussed in the next section.

Now our Django project is configured to support caching infrastructure using Redis. Django, out of the box, supports caching responses and different utilities. DRF uses the utilities provided by Django to support caching responses and different cache functionalities.

> **Read more**
>
> Django's caching framework provides the basic caching feature that can cache responses. The official documentation has all the details to implement view-level caching. We are not going to deep dive into them since this caching interface provides a basic implementation that most developers are already aware of.
>
> Django: `https://docs.djangoproject.com/en/stable/topics/cache/#the-per-view-cache`
>
> DRF: `https://www.django-rest-framework.org/api-guide/caching/`

Apart from the view-level caching interface, Django also provides a low-level API to implement our own caching interface. We have learned in our previous chapters about how we can utilize DRF to build modern-day applications, and caching is an important piece in this. We can use low-level APIs that Django provides to create our caching layer, but we have to write a lot of additional code, which is not recommended. Hence, I recommend using an additional package: `django-cacheops`.

Using django-cacheops

Django 4 provides out-of-box support for Redis as a caching layer, but the default caching interface provided by Django is more suitable for caching views. As we build complex features in our applications we would need to implement caching on business logic implemented inside functions. Django provides a low-level API interface to use the underlying caching framework and work directly with

Redis, but it becomes too much of a hassle to implement caching functionality for developers. For this purpose, I always recommend using django-cacheops (https://github.com/Suor/django-cacheops), a third-party package that abstracts out all the additional overhead working with low-level caching API. Here is what the django-cacheops package says:

Django cacheops is a slick ORM cache with automatic granular event-driven invalidation.

Let us quickly see how we can configure Redis with django-cacheops. First, add cacheops to the INSTALLED_APPS list in the settings.py file. Now use the same configuration of the Redis server we got from the Redis labs in the previous section:

```
CACHEOPS_REDIS = {
    "host": "<public endpoint>",  # Redis endpoint
    "port": 15014,  # for redis lab, port is 15014
    "socket_timeout": 3,  # connection timeout in seconds, optional
    "password": "<REDIS_PASSWORD>",
}
```

Now that we have configured django-cacheops in our project, let us implement the most common and widely used caching use case. Let us try to get all the blogs written by a given author:

```
@cached(timeout=60*10)
def get_all_blogs(author_id):
    print('Fetching blogs from database')
    blogs = Blog.objects.filter(author_id=author_id)
    blogs_data = BlogSerializer(blogs, many=True).data
    return blogs_data


@api_view(['GET'])
def get_blogs_by_author(request):
    author_id = request.GET.get('author_id')
    blogs = get_all_blogs(author_id)
    return Response({'blogs': blogs})
```

In our example, we are caching the response of get_all_blogs for different author_id for 10 minutes. This means that if we pass the same author_id to the function for 10 minutes, the output will be returned using the cached result saved in Redis. The caching is done by the cached decorator. To verify our caching, we have added a print statement. Now, for the first time, we should see the print statement output in our terminal: Fetching blogs from database.

The cached decorator takes timeout as an argument that signifies the total time in seconds, after which the cached value would be removed from Redis. cacheops also takes care of creating unique caching key names and other low-level implementation details that are important to keep in mind to make sure there is no conflict.

> **Important note**
>
> One should make sure the returned value is an evaluated QuerySet with values and not a QuerySet before making a DB query. In our example, we have a `return list(blogs)`, which means we are returning the value of the query; if we had returned the `blogs` variable, then the caching would work differently. Though the function would have been cached, it would return a cached QuerySet that would always hit the database to return the value.

We need to also implement a way to invalidate the cache so that we do not use stale data from the cache after we have updated the data in the database. `django-cacheops` provides a simple `invalidate('<key>')` interface to remove all the cached results. For example, if we have to invalidate the result for author ID 12, then we need to call:

```
get_all_blogs.invalidate(12)
```

If we have to invalidate all the cached results for the given function, do not pass any argument while calling the invalidate. For example, to remove all the cached results for the `get_all_blogs` function, we need to call invalidate without any argument:

```
get_all_blogs.invalidate() # removes all cached response
```

Let us now list all the features and caveats that one should know while working with `django-cacheops`:

- You should use a simple time-invalidated cache to optimize costly DB queries:

    ```
    @cached(timeout=no_of_sec)
    def some_costly_function(data_input):
        return …
    ```

    Automatically, our function is now cached for a particular `data_input` during the given time frame. For different values of `data_input`, we shall have different results computed and cached. The `timeout` argument takes input in total seconds, after which the cached result will be evicted from Redis.

- Implement a manual invalidation interface that can easily be accessed when needed:

    ```
    some_costly_function.invalidate('abc')
    # This would invalidate the caching for the input argument of
    # 'abc' but would have the other functions cached

    some_costly_function.invalidate()
    # This would invalidate the caching for overall function.
    ```

- Implementing caching on auth tables is a double-edged sword; one should avoid it unless there is strong latency associated with DataBase query for auth tables.

- Cacheops provides the interface to cache all tables for Django apps using the configuration in the settings.py file. In my opinion, this is something one should avoid because it can cause a lot of confusion for developers unable to see expected results in production since there is a blanket caching layer on the model table. I always recommend using function-level caching, as this is more explicit and verbose for anyone to detect.

- If there is a strong need to implement caching in the DB layer—for example, if we have to cache all the get queries for auth user table—we can configure cacheops in the Django settings.py file:

```
CACHEOPS = {
    # Cache any User.objects.get() calls for 15 minutes
    # This also includes .first() and .last() calls,
    # as well as request.user or post.author access,
    # where Blog.author is a foreign key to auth.User
    'auth.user': {'ops': 'get', 'timeout': 60*15},

    # Cache all gets and QuerySet fetches
    # to other django.contrib.auth models for an hour
    'auth.*': {'ops': {'fetch', 'get'}, 'timeout': 60*60}
}
```

- Use cached_as decorator whenever you want to implement the auto invalidation of cached results on any data source update. For example, if we want to invalidate cache on any blog table update, we should use the cached_as decorator and pass the model name Blog as a parameter:

```
from cacheops import cached_as
@cached_as(models.Blog, timeout=15*60)
def get_all_blogs(author_id):
    blogs = models.Blog.objects.filter(author_id=author_id)
    return list(blogs)
```

  Whenever there would be any update in the Blog model data, then the cache would be automatically invalidated in the get_all_blogs function.

Apart from the mentioned tips and caveats, it is highly recommended to read all the additional information provided in the official documentation. Now let us learn a few good practices to be followed while implementing caching in production.

Best practices for caching in production

Let us see a few important points we should keep in mind while implementing caching in production:

- Always use high-availability Redis cluster on production. It is recommended to use managed services such as ElastiCache by AWS or Redis Lab with master–slave configuration for production. This becomes important because if you have a high read system and caching goes down, all the read queries hit the database, which can create a cascading effect and bring the whole system down.

- Avoid premature caching on every DB query. A lot of developers try to save an additional 10 ms of database query calls and make the system complicated.

- Always be verbose on the caching implementation and add caching on functions rather than overall views or DB tables.

- Cache invalidation is an important part of cache implementation. Always create interfaces to invalidate the cache in production. I have personally observed a couple of production issues caused by the lack of a cache invalidation mechanism; for example, a particular data was updated in the database, but users were seeing stale data. So, as a thumb rule, if you are caching a few important functions, create an invalidation interface early on to avoid any production incident.

- Never cache any API or function that deals with sensitive data. For example, never cache user authentication tokens. Caching authentication tokens should not be done because we want to perform authentication each time.

We have learned about caching and how we can implement caching in Django, DRF, and using `django-cacheops`. In the following section, we will learn about the concept of throttling and how DRF provides us with the interface to implement throttling in APIs built in Django. Throttling uses the same default caching infrastructure we mention in the Django setting configuration.

Throttling with Django

Throttling is the process of limiting the number of API requests that can be made by a user in a given time frame. Django does not provide any throttling feature out of the box, but DRF does. In this section, we will learn about how we can incorporate throttling into using DRF.

Different systems have different use cases for throttling. For a SaaS product, throttling is needed to limit the number of API calls per customer per minute depending upon their paid plan. An example is Shopify backend APIs. A standard product company doesn't want its APIs to be abused by any outside customer, so it would throttle any request beyond a particular threshold.

> **Important note**
> We should understand that in this section we are implementing throttling in the application layer. This is generally not a recommended approach to protecting your service from brute force attacks since application-level throttling might still increase the request queue for your application and degrade the overall system performance. We should always implement throttling using a **web application firewall** (**WAF**). This will protect your system from queuing up requests when there is a large brute-force attack. Do note that when you are building a new system and not expecting high traffic, you can postpone the incorporation of WAF in your system for a later point in time. Application-level throttling provided by DRF should be used to incorporate business tier implementations.

DRF uses Django's default caching framework as a backend. Before setting up throttling, we should set up caching in Django; we do not have to do anything extra apart from the basic Django Redis setup we learned in the previous section. Once we have Redis set up with Django backend caching, we need to configure the throttling classes. It is recommended not to use blanket throttling via `DEFAULT_THROTTLE_CLASSES` configuration for all the APIs using `setting.py` because different APIs would have different throughputs. It is better to create an individual throttle limit after assessing traffic patterns for each endpoint independently. We need to define the default throttle rate in the `settings.py` file. The `'anon'` key represents how many *non-logged-in requests* are allowed and the `'user'` key represents the configuration for the *logged-in requests*:

```
REST_FRAMEWORK = {
    'DEFAULT_THROTTLE_RATES': {
        'anon': '100/day',
        'user': '1000/day'
        'scope': '10000/day'
    }
}
```

DRF provides three different out-of-the-box throttling classes that can help us implement throttling:

- `AnonRateThrottle`: Implement throttling for non-logged-in user requests. By default, the throttling rate would be taken as the value passed in the `'anon'` default throttle rate config. Internally, DRF would use the `X-Forwarded-For` header to identify the IP address and determine the different sources of the request. Hence, it is very important to configure the application to forward the IP address of the client in the given header.

- `UserRateThrottle`: Implement throttling for logged-in user requests. By default, the throttling rate would be taken as the value passed in the `'user'` default throttle rate config. This puts a global rate limit on the total number of APIs called by the end user; hence, one should be cautious while attaching `UserRateThrottling`. It would use the user ID to generate the unique throttle key. If not found, it would fall back to the IP address. The default scope `'user'` key can be overridden by assigning the new scope value to the extended class.

- `ScopedRateThrottle`: Implement throttling to specific APIs clubbed together. For example, if we want to implement throttling for two APIs combined, we should use `ScopedRateThrottle` with a defined scope.

DRF APIViews has `throttle_classes` as a property that takes a list of all the throttle classes. If we have to attach an `AnonRateThrottle` to `BlogApiView`, we can do it by adding the throttle class in the following code snippet:

```
from rest_framework.views import APIView
from rest_framework.throttling import AnonRateThrottle
```

```
class BlogApiView(APIView):
    throttle_classes = [AnonRateThrottle]
    ...
```

Similarly, we can attach the `UserRateThrottle` class to `APIView`. For `ScopedRateThrottle`, we have to pass an additional `throttle_scope` property. For example, `BlogApiView` and `BlogDetailApiView` both have the same `throttle_scope` assigned, which means the total rate limit is going to be distributed among both the APIs while `Blog2ApiView` shall have an independent limit:

```
from rest_framework.views import APIView
from rest_framework.throttling import AnonRateThrottle

class BlogApiView(APIView):
    throttle_classes = [ScopedRateThrottle]
    throttle_scope = 'blog_limit'
    ...
class BlogDetailApiView(APIView):
    throttle_classes = [ScopedRateThrottle]
    throttle_scope = 'blog_limit'
    ...
class Blog2ApiView(APIView):
    throttle_classes = [ScopedRateThrottle]
    throttle_scope = 'blog_2_limit'
    ...
```

We can set the throttling limit in the `settings.py` file, as shown in the following snippet:

```
REST_FRAMEWORK = {
    'DEFAULT_THROTTLE_RATES': {
        'blog_limit': '1000/day',
        'blog_2_limit': '100/day'
    }
}
```

All the throttle classes are expandable and one can easily override them. It is also possible to pass multiple throttle classes to one APIView, so we can add multiple rules for a single API view. If there are multiple throttle classes attached to a single APIView, all the classes need to pass before executing the main body of the view or else `exceptions.Throttled` is raised.

For more details on how we can create custom throttle classes and extend existing classes, you can look into the official documentation: `https://www.django-rest-framework.org/api-guide/throttling/`

Best practices for throttling in production

Now let us look into a few best practices one should follow while implementing throttling in production:

- Always prefer to implement throttling using WAFs or other components. Application-level throttling should be implemented only if the application is small and the added complexity of using other methods outweighs the advantage.

- Avoid adding blanket application throttling on DRF settings. It can backfire at any given time on production.

- Add monitoring and logging for all the throttled requests. This helps you monitor any unexpected throttling behavior or increase the throttling limit from time to time as per the usage pattern.

- Always remember to pass the correct IP address of the client to `X-Forwarded-For` so that the throttling backend performs the expected behavior. As we have multiple components between the client and the Django server, it is common to use the wrong IP address. Always double-check that the client IP is propagated properly to the Django server.

- The default `UserRateThrottle` always uses a global rate limit. It is recommended to use a custom scope by extending the `UserRateThrottle` class.

We shall now learn how we can implement logging in Django and how logging can help us function better in production.

Logging with Django

Logging can be considered a record of every data in programming. Django uses Python's built-in logging module (`https://docs.python.org/3/library/logging.html`) to capture logs. In Django, logging is configured as part of the general Django `django.setup()` function, so it's always available unless explicitly disabled. Python's logging module provides extensive options to set up and configure logging; hence, all those configurations are also available when we use logging in Django. In this section, we shall learn how we can set up logging in Django and the best practices to use logging in Django projects.

> **Important note**
>
> Logging in Python is an extensive topic, and the configuration of logging can be tricky at times. In this section, we will not go into too much detail about how one can use all the different options provided by Python. Rather, we shall check the standard configuration, which I have used in multiple projects and have found to be super helpful.
>
> If you want to deep-dive into detailed configurations, please go through the Python logging module official documentation and Django logging documentation: `https://docs.djangoproject.com/en/stable/topics/logging/`

Setting up logging

By default, Django logs are sent to the console. Whenever we add a print statement to our Django code, the output is printed out in the console. While working on the local setup, this is the first approach taken by developers to log data and develop the application. However, this approach doesn't work in production since developers don't have access to the console. This is where logging output to files is helpful.

Let us see how we can configure Django to set up the logging infrastructure. Please note that we are going to work with just one standard configuration that mostly solves all the standard use cases of logging. To enable logging in Django, we need to add a snippet to the Django `settings.py` file, as seen in the following code snippet:

```python
LOGGING = {
    "version": 1,
    "disable_existing_loggers": False,
    "formatters": {"verbose": {"format": "%(asctime)s %(process)d
%(thread)d %(message)s"}},
    "loggers": {
        "django_default": {
            "handlers": ["django_file"],
            "level": "INFO",
        },
    },
    "handlers": {
        "django_file": {
            "class": "logging.handlers.RotatingFileHandler",
            "filename": "path/to/django_logs.log",
            "maxBytes": 1024 * 1024 * 10,   # 10MB
            "backupCount": 10,
            "formatter": "verbose"
        },
    },
}
```

Let me quickly explain what each line in the LOGGING object signifies:

- `version`: This means the configuration version used is `logging.config.dictConfig` *version 1* format.
- `disable_existing_loggers`: Django provides a few default logging configurations. The `disable_existing_loggers` flag is used to enable/disable the default logging configuration setting. In general, keep this flag false. This means the default loggers configured by Django are still active.
- `formatters`: Formatters are key–value pairs, with each key representing a formatter we can use for logging messages. We have configured a `verbose` formatter to log additional information such as timestamps, process IDs, etc.

- `loggers`: Loggers are namespaces used when we create logs. One application can have multiple loggers. In our application, we are naming the custom logger `django_default`, and whenever we have to use this logging configuration, we would use this name:

  - `level`: Levels are used to filter out log levels. Here, we have defined the `INFO` level, which means any log level that is equal to or above the `INFO` level would be using this configuration.

  - `handlers`: Handlers are used to map each logger namespace to different handlers. These handlers are engines that determine how to process each log message. One can have multiple handlers associated with a logger. In this case, we have named the handler `django_file`, which we, again, defined and configured in the subsequent section.

- `handlers`: Handlers are the engine to logs, they determine how to process each log message. Each log message can write a message to the console, send an email, write to a log file, or pass it to the network via sockets. Handlers specify how to propagate these messages. In our config, we are defining the `django_file` handler that writes logs to files. Let us look into each key–value pair we have defined in the handler:

  - `class`: This determines how the log messages are going to be processed. We are passing the `RotatingFileHandler` class, which means new log files are going to be created every time we have our previous log file filled with 10 MB logs that we are configuring. We could also simply use `logging.FileHandler` if we don't want to rotate new files.

  - `filename`: This defines the name of the file in which the log message content would be saved. We should always provide an absolute path for the file.

  - `maxBytes`: This is specific to rotating log files. It determines the maximum size of the file before a new file should be created.

  - `backupCount`: This is the total number of file counts that would be created before reusing the older files by rotation. The file names would have the file count number added to the end.

  - `formatter`: This is the formatter we want to use to save the messages received in the handler.

The configuration we discussed is the bare minimum that you can use to work on the production. You can have a different use case and may add more configurations to this boilerplate logger config. If you are just starting to learn about loggers in Django, I would recommend that you stick to the mentioned configuration and evolve as your requirements grow.

Now we have set up logging in Django. Passing log messages and writing them to log files is a straightforward job. But if you don't have a centralized function to pass logs to as your application matures, the captured logs would become difficult to read and a lot of inconsistency would be introduced. I recommend creating a common `logger` function and passing log messages to the `logger` function. We have created a `log_event` function that can be called from different functions in our Django project. Create a new file in the `common/logging_util.py` with the following code:

```
import json
import logging
```

```
from common.localthread_middleware import get_current_user_id
from common.localthread_middleware import get_txid

def log_event(event_name, log_data, logging_module="django_default",
level="INFO"):
    """

    :param event_name: Event name which you are logging
    :param log_data: The data you want to log, this can be anything
serializable
    :param logging_module: If you want to use any custom module for
logging, define it in Django settings
    :param level: Level for which you are logging.
    """
    logger = logging.getLogger(logging_module)

    try:
        msg = {"ev": event_name, "data": log_data, "txid": get_txid()}
        user_id = get_current_user_id()
        if user_id:
            msg["uid"] = user_id
        logger.log(msg=json.dumps(msg), level=getattr(logging, level))
    except Exception as e:
        print('Error')  # use error monitoring tool
        return
```

In our snippet, we have the get_current_user_id function, which returns the current user ID from the request object. We need to use a custom middleware to get the current user ID. We shall discuss custom middleware in detail in *Chapter 7*.

In this section, we shall only use the custom middleware without getting into the concepts. Create a new file common/localthread_middleware.py with the following code:

```
from threading import local
from uuid import uuid4

_thread_locals = local()

def get_current_request():
    """Returns the request object in thread local storage."""
    return getattr(_thread_locals, "request", None)

def get_current_user():
    """Returns the current user, if exist, otherwise None."""
    request = get_current_request()
```

```
        if request:
            return getattr(request, "user", None)

def get_txid():
    """Returns the current transaction id, else None."""
    return getattr(_thread_locals, "txid", None)

def get_current_user_id():
    """Returns authenticated user id, else returns 0."""
    user = get_current_user()
    if user and user.id:
        return user.id
    return 0


class PopulateLocalsThreadMiddleware:
    def __init__(self, get_response):
        self.get_response = get_response


    def __call__(self, request):
        print("custom middleware before next middleware/view")
        # Populate the request object in thread local storage to be
accessible anywhere
        _thread_locals.request = request
        _thread_locals.txid = str(uuid4())
        # the view (and later middleware) are called.
        response = self.get_response(request)
        # Clean up the thread local storage
        _thread_locals.request = None
        print("custom middleware after response is returned")
        return response
```

We shall discuss the custom middleware code in detail in *Chapter 7*. For now, we will focus on the logging functionality. Now, we need to link the custom middleware in our `settings.py` file to link to our Django application:

```
MIDDLEWARE = [
    ...
    'common.localthread_middleware.PopulateLocalsThreadMiddleware',
]
```

Now we have successfully created a `log_event` function that can be used to capture log messages and processes them to the required format. Our `log_event` function would also automatically populate the `user_id` for each log message. We can also create a unique `transaction_id` for each request and use them to track logs across the stack. Let us discuss the `log_event` function and the information added:

- `event_name`: The event name is useful when searching through thousands of logs. We pass the event name to the `ev` key.

- `log_data`: This is serializable data that contains the log message. This can be either a string, dictionary, or list.

- `logging_module`: By default, we are picking the `django_default` logger we defined earlier.

- `level`: `level` defines the type of logs we want to create using the function. By default, we are using the `INFO` level.

- `uid`: This is the user ID extracted from the request.

- `txid`: This is the unique transaction ID for each request.

Apart from these, we automatically save the current timestamp and other required information in the log messages. We need to call the `log_event` function and pass the required information to start logging. For example, if we want to log the `author_id` each time someone makes a new request, we can use the following code:

```
from common.logging_util import log_event


def get_blogs_by_author(request):
    author_id = request.GET.get('author_id')
    log_event('get_blogs_by_author', {'author_id': author_id})
    blogs = get_all_blogs(author_id)
    return Response({'blogs': blogs})
```

Whenever we make a request, we can see log files getting populated by the request. *Figure 6.2* shows the log file content that was generated by the `log_event` function:

```
1   2023-12-03 08:30:57,014 67573 6195982336 {"ev": "get_blogs_by_author", "data": {"author_id": "1"}, "tx_id": "b8e74a0c-ed34-41b1-9cd7-0c2fc9c1bc73", "uid": 1}
2   2023-12-03 08:32:01,204 67609 6130790400 {"ev": "get_blogs_by_author", "data": {"author_id": "1"}, "tx_id": "874e4129-6ced-4920-bafc-390a5eb4a7a8", "uid": 1}
3   2023-12-03 08:32:01,205 67609 6130790400 {"ev": "demo", "data": {"author_id": 1}, "tx_id": "874e4129-6ced-4920-bafc-390a5eb4a7a8", "uid": 1}
4   2023-12-03 08:40:33,611 67776 6200078336 {"ev": "get_blogs_by_author", "data": {"author_id": "10"}, "txid": "b763ac38-0828-463a-be03-9fac70b93ea7", "uid": 1}
5   2023-12-03 08:40:33,612 67776 6200078336 {"ev": "demo", "data": {"author_id": 1}, "txid": "b763ac38-0828-463a-be03-9fac70b93ea7", "uid": 1}
6   2023-12-03 08:40:36,545 67776 6200078336 {"ev": "get_blogs_by_author", "data": {"author_id": "12"}, "txid": "de083545-2ae1-468a-82b0-93031e171d34", "uid": 1}
7   2023-12-03 08:40:36,547 67776 6200078336 {"ev": "demo", "data": {"author_id": 1}, "txid": "de083545-2ae1-468a-82b0-93031e171d34", "uid": 1}
8   2023-12-03 08:40:38,876 67776 6200078336 {"ev": "get_blogs_by_author", "data": {"author_id": "13"}, "txid": "6221a683-f0f1-45f3-b0d5-817aad2a93d9", "uid": 1}
9   2023-12-03 08:40:38,877 67776 6200078336 {"ev": "demo", "data": {"author_id": 1}, "txid": "6221a683-f0f1-45f3-b0d5-817aad2a93d9", "uid": 1}
10  2023-12-03 08:40:40,889 67776 6200078336 {"ev": "get_blogs_by_author", "data": {"author_id": "12"}, "txid": "4df0b0e2-0ea5-4447-b4a9-3b32f3e64d8e", "uid": 1}
11  2023-12-03 08:40:40,889 67776 6200078336 {"ev": "demo", "data": {"author_id": 1}, "txid": "4df0b0e2-0ea5-4447-b4a9-3b32f3e64d8e", "uid": 1}
12  2023-12-03 08:40:46,155 67776 6200078336 {"ev": "get_blogs_by_author", "data": {"author_id": "49"}, "txid": "fb7a9916-fd5a-4bd5-8b11-2906c823470a", "uid": 1}
13  2023-12-03 08:40:46,155 67776 6200078336 {"ev": "demo", "data": {"author_id": 1}, "txid": "fb7a9916-fd5a-4bd5-8b11-2906c823470a", "uid": 1}
```

Figure 6.2: Log file content

We might want to manually open each file and read the log messages, but this is not a scalable approach when we have multiple servers. We can use different log processing stacks to perform this. One of the most common and widely used stacks is **Elasticsearch-Logstash-Kibana (ELK)**, but this needs some manual effort to set up. I recommend using **NewRelic** to process and navigate through logs. We will learn about the setup of NewRelic and how to use NewRelic to process and access logs in *Chapter 14*. Let us learn a few best practices we should follow for logging in production

Best practices for logging in production

We have learned how we can set up logging in Django and how we can create and access log messages. Now let us learn a few best practices to consider while using logging in production:

- Do not use logging to collect errors on production. One of the most common mistakes we see is developers using logging to capture errors on production. Though this works for small projects, this approach is not scalable. One should use error monitoring tools such as **Sentry** or **Rollbar** for this purpose. We shall learn more about these tools in *Chapter 14*.

- Do not use emails to transfer logs. Rather, use log management agents to transfer generated logs to a log engine to have better accessibility to logs. Use tools such as NewRelic or ELK stack to access and navigate through logs.

- Avoid passing sensitive information to logs. Our application might have access to sensitive information about the user, and as we start using logs to capture different information, we should make sure we are not capturing sensitive information in our logs. For example, avoid capturing email addresses, passwords, payment details, etc. This can be achieved by using custom logging filters that the Python logging module provides.

- Use the common logging function, as shown earlier. This will give the logs a consistent format, make them easy to parse by logging agents, and help us scan through them better.

- Remove logs after the current use case is over. If we do not remove all the old and unnecessary logs, this might incur high cost due to large volume of log consumption. As your application grows, you should track how you can remove old logging function calls.

We have learned how to implement logging in Django and the best practices for implementing logging in our project. This can be considered as a boilerplate setup for logging into Django. Most of the information using logging can be gained through the mentioned setup. However, if you have different requirements, feel free to access the official documentation and improvise your setup.

Summary

In this chapter, we learned how to set up a Redis server using RedisLabs. Redis is an important in-memory database used for different applications such as caching and throttling. We learned how to integrate Redis into native Django and use caching with Redis. The built-in support for caching in Django is not sufficient to add caching to APIs, but `django-cacheops`, a third-party package, can ease the caching implementation. We also learned how to set up `django-cacheops` with Redis and the best practices for using it in production.

Apart from caching, Redis is used for throttling, and we have learned how to set up throttling in Django and how to create custom application-level throttling. Logging in Django uses Python's built-in logging module. It is a daunting task for developers to set up logging in Django, so we learned about the basic boilerplate logging setup that you can use in production to log messages. We will revisit logging setup again in *Chapter 14*, where we will learn how to set up logging with NewRelic and ease the overall experience of navigating through logs.

In *Chapter 7*, we will learn how to use Pagination in DRF and Django Signals. We will also discover how we can create custom middleware in Django and different applications of middleware.

7

Using Pagination, Django Signals, and Custom Middleware

In *Chapter 6*, we learned how to integrate Redis into the Django project and utilize Redis as a caching layer. Caching is important to get performance gains, but at times, we would want to optimize the overall response sent to the client. This can be achieved by implementing **pagination**.

In this chapter, we will learn how to implement pagination in Django and **Django Rest Framework** (**DRF**). Along with that, we will focus on Django signals and how they help developers decouple logic between different apps. In *Chapter 5*, we saw how Django authentication and authorization support using Django middleware out of the box; now we will learn how to create custom middleware in Django.

We will cover these topics in this chapter:

- Paginating responses in Django and DRF
- Demystifying Django signals
- Working with Django middleware
- Creating custom middleware

Technical requirements

In this chapter, we will learn how to implement pagination in Django and DRF. We expect you to be familiar with the term *pagination* and how it is advantageous. Django signals and custom middleware are also advanced topics, so we expect you to be familiar with all the concepts we have covered in the previous chapters. You are expected to be familiar with the Django request-response cycle, which we discussed in *Chapter 3*.

Paginating responses in Django and DRF

When a request is sent by the client to fetch records from the server, it is not recommended to send all the entries present in the database in one go. This is when the concept of pagination comes into the picture. For example, if we have 1,000 blogs saved, sending all the blogs at once in response to the client's request would increase the page load time and add performance bottlenecks to the system. To solve these problems, we will introduce the concept of pagination.

Understanding pagination

Let us take an example to understand the concept of pagination. A client requests to get a list of blogs, and we have 1,000+ blogs published on our website. We would send a few blogs in small batches, say 10 blogs every time. This is known as pagination. The user would interact with the UI and request the next batch of blogs. Different types of pagination can be implemented by the server for the API:

- Cursor API pagination
- Keyset API pagination
- Offset API pagination
- Time-based API pagination

We will learn about all these techniques in the *Using pagination in DRF* section. In short, we can consider pagination to be the concept where we have a list of entries and we slice the list into smaller sections before sending each slice of data to the client.

Let us learn about the basic pagination that can be implemented in the simplest way using Django ORM by creating a `get_blog_without_pagination` view that returns all the blogs. This view would work for a small scale, but when we have 1,000+ blogs, then this view would be slow and not very useful since readers are not expected to read all 1,000+ blogs. Add the following code to `blog/views.py`:

```
def get_blog_without_pagination(request):
    blogs = Blog.objects.all()
    blogs_data = BlogSerializer(blogs, many=True).data
    return Response({'blogs': blogs_data})
```

We also need to link our view to the URLs, so add the following code to the `blog/urls.py` file:

```
urlpatterns = [
    . . .
    path('unpaginated/', views.get_blog_without_pagination),
    . . .
]
```

When we open `http://127.0.0.1/blog/unpaginated/`, we will get the list of all the blogs created in our database. This is not scalable as we have more blogs created.

In order to add pagination, we need to break down the response into smaller chunks, say 10 blogs per response. We can do it using the limit offset of Django ORM. Let us implement pagination for the same view. Use the following code to create a new `get_blog_with_pagination` view:

```
@api_view(['GET'])
def get_blog_with_pagination(request):
    page = int(request.GET.get('page', 1))
    page_size = int(request.GET.get('page_size', 10))
    offset = (page-1)*page_size
    limit = page*page_size
    blogs = Blog.objects.all()[offset:limit]
    blogs_data = BlogSerializer(blogs, many=True).data
    return Response({'blogs': blogs_data})
```

Also, we need to link our view to the URLs by adding the following code to the `blog/urls.py` file:

```
urlpatterns = [
    . . .
    path('paginated/', views.get_blog_with_pagination),
    . . .
]
```

When we open `http://127.0.0.1/blog/paginated/?page=1&page_size=2`, we will get the list of two blogs created. Now the user requests blog data with the query parameter and passes the page number for which blogs need to be returned. Our API call would take `page` and `page_size` as input and return the number of blogs accordingly.

While building product features with Django, we need to implement pagination. For almost all use cases, we can use custom pagination logic, as discussed in this section.

DRF also provides an interface to use different pagination techniques out of the box. Let us learn how we can use pagination using DRF.

Using pagination in DRF

Django provides the `Paginator` class (`https://docs.djangoproject.com/en/stable/topics/pagination/`) out of the box, which can be used to implement pagination. DRF makes use of the same `Paginator` class to provide pagination support to APIs. We will not get into the details of the `Paginator` class provided by Django; rather, we will focus on what type of pagination is provided by DRF.

Please note

We are not discussing the integration of pagination in DRF since the documentation of DRF provides details of how you can set up different types of pagination. You can follow the implementation details from the official documentation (`https://www.django-rest-framework.org/api-guide/pagination/`).

In our GitHub code repository, we have added different examples to integrate pagination for DRF. Please check the README file of `Chapter07` for more details: `https://github.com/PacktPublishing/Django-in-Production/tree/main/Chapter07`.

There are primary three types of pagination available:

- `PageNumberPagination` – This pagination sends the page number in the URL query parameter. For example, if we have 100 blogs and we are splitting the blogs into batches of 20, then we will have 5 pages of 20 blogs each. Whenever the client requests page 1, we will provide the first 20 blogs; as the page number in the client request increases, we will share the next batch of blogs. Here, the client doesn't have control over how many entries will be fetched in the response.

- `LimitOffsetPagination` – Here the client has exact control over how many entries need to be fetched in the response and the offset from where the data can be fetched. This type of pagination gives the client more control over the data retrieval but can also introduce vulnerabilities to the API. For example, a malicious user could hit the API with a limit of 1,000 and this would put unnecessary pressure on our server. Hence, to solve this problem, we should put a check that would stop any unrealistic `limit` value passed in the client request that could put a load on the server.

- `CursorPagination` – Cursor-based pagination is a bit of a complicated pagination technique that needs some attention to detail while implementing else it can easily backfire. Cursor-based pagination is a scalable technique to implement pagination in large-scale systems. I would recommend developers refrain from implementing cursor-based pagination until there is a strong scaling problem involved that the other methods are unable to address. For more details on the concept, please read the article by David Carmer on how Disqus implemented cursor-based pagination (`https://cra.mr/2011/03/08/building-cursors-for-the-disqus-api/`).

Just like other DRF classes, you can easily extend the built-in pagination classes to create custom `Pagination` classes as per the use case.

In this section, we have learned how pagination is useful to improve the performance of a website and also how to work with different pagination classes in DRF. Now, let us look into another advanced concept that Django provides: Django signals.

Demystifying Django signals

Django has a lot of out-of-the-box features, but one feature that is underrated and stands out is Django signals. Django includes a "signal dispatcher" out of the box that helps developers write decoupled logic. A signal dispatcher would notify a set of receivers that some action has taken place. This is useful when we want to execute multiple actions for a single event. By default, Django provides a set of built-in signals that are useful for building common applications, as follows:

- **Model signals** – These are database action signals that can help developers trigger actions whenever some kind of database operation is about to be performed or is already being performed. Here is the list of default Django signals that are dispatched for certain Django model operation events:

  - `pre_save` and `post_save` – These signals are triggered before and after the model's `save()` method is called. Please note that these signals are *not triggered* when the `update()` method is used.

  - `pre_delete` and `post_delete` – These signals are triggered before the `delete()` method is called from the `Model` object or query set.

  - `m2m_changed` – This signal is triggered when a `ManyToManyField` of Django model is changed. Since all the previous signals mentioned in this list until this point are not applications for M2M fields, `m2m_changed` is triggered whenever there is any model change. The arguments sent by `m2m_changed` have `action` data that can have either of the following as its value:

    - `pre_add` and `post_add` – Sent before and after one or more objects are added to the relationship of the many to many field.

    - `pre_remove` and `post_remove` – Sent before and after one or more objects are removed from the relationship of the many to many field.

    - `pre_clear` and `post_clear` – Sent before and after the M2M relation is cleared of the many to many field.

- **Management signals** – These signals are fired when the `manage.py migrate` command is executed. One of the use cases I have seen is to use it to notify the developer team that a new migration change is executed on production. Here is the list of default Django signals that are dispatched for certain Django management command executions:

  - `pre_migrate` – Fired when the migration process starts

  - `post_migrate` – Fired after the migration process is complete

- **Request/response signals** – These signals are fired when a request is received by the Django application and when the response is processed and sent back to the client. Here is the list of default Django signals that are dispatched for Django request/response events:

  - `request_started` – Sent when Django receives the signal from the client and starts processing

  - `request_finished` – Sent when Django delivers the HTTP response to the client

  - `got_request_exception` – Triggered whenever Django encounters an exception in the request/response cycle

- **Database wrappers** – Whenever Django connects to the database, these signals are triggered. It is helpful to perform any custom operation on a Django database connection. One of the common use cases of setting a query timeout in a database is discussed in *Chapter 2*. The following Django signal is dispatched when we have the DB connection established:

  - `connection_created` – This is fired whenever a database connection is established between the Django server and the database

We have learned about the default signals Django provides to us that can be used. As a project grows, the requirement for signals also evolves. Hence, sometimes the signals we just discussed might not be sufficient to support all use cases. Django provides a way to create custom signals, as discussed in the following section.

Creating custom signals

Django signals are a powerful mechanism to write decoupled logic inside a Django project. The built-in signals mentioned in the *Demystifying Django signals* section help us with default use cases, but if we have a custom use case, Django custom signals can help us address it. For example, if we want to send multiple communications such as email, SMS, and WhatsApp whenever a user purchases a product on our website, we can do it via custom Django signals. We would dispatch a custom signal on every purchase made and our receivers would listen to the custom signal and act on every new signal received.

Let us now try to implement a custom signal that would always send out an email whenever a new blog is published:

1. Create a `signals.py` file (this can be named anything, but it is preferable to name it `signals.py` for easier identification) inside the `blog` app. Add the definition of the signal in the `blog/signals.py` file:

```
from django import dispatch
notify_author = dispatch.Signal()
```

2. We want to dispatch the `blog_published` signal whenever we publish the blog. We will add the trigger on the `publish_blog` method in the `blogs/public.py` file:

```
from blogs import signals
def publish_blog():
    # publish blog logic
    signals.notify_author.send(sender=None, blog_id=123)
```

3. We have now dispatched the signal whenever a blog is published. We need to add a receiver to our application that will listen to the dispatched signal and perform actions. Create an `author/receiver.py` file in the `author` app:

```
from django.dispatch import receiver
from blog import signals
@receiver(signals.notify_author)
def send_email_to_author(sender, blog_id, **kwargs):
    # Sends email to author
    print('sending email to author logic', blog_id)
```

4. Now, we need to make sure the receiver is registered when the app is loaded. This is the most important step. (*I have lost many developer hours debugging why receivers didn't work.*) The following code snippet is required to plug in Django signals to a Django app. Go to the `author/apps.py` file and add the following code:

```
from django.apps import AppConfig
class ApplicationConfig(AppConfig):
    name = "author"
    def ready(self):
        from author import receivers
```

That's it! Now you have a custom signal that will send emails whenever a new blog is published. Though the implementation is fairly simple, the tricky part is making sure that all the steps are implemented properly.

5. To test our signal, let us connect our `blog/public.py` code with a view and access the endpoint. Create a view to link the signal code by adding the following code to the `blog/public.py` file:

```
@api_view(['GET'])
def publish_blog(request):
    blog_id = request.GET.get('id')
    public.publish_blog(blog_id)
    return Response({'status': 'success'})
```

We need to connect the view to the `blog/urls.py` file.

```
urlpatterns = [
    path('publish/', views.publish_blog)
]
```

Now, when we go to the `http://127.0.0.1/blog/publish` endpoint in our browser, we will see the signals working perfectly. Go to the terminal and check for the message `Sending email to the author logic` to validate the signals integration.

By default, Django signals are synchronous in nature. If your task is going to interact with external systems and can have a higher expected latency, it is always favorable to create asynchronous signals with Celery. We will discuss Celery and how to create asynchronous signals in *Chapter 8*.

Let us now check a few best practices for using signals in Django in production.

Working with signals in production

Django signals are considered to be an advanced feature. Most of the tutorials available online don't cover how you should use Django in production or discuss different patterns. In this section, we will learn all the important points you should keep in mind while working with Django signals in production, which are as follows:

- Avoid using signals if you do not have a strong use case. Yes, always try to avoid using signals because, as the project grows, developers struggle to debug code connected via signals.

- Create a separate file (say, `signals.py`) that will have all the signals for the given Django app in one place. It might be tempting to use a `models.py` file to add all the signals, but this becomes chaotic as the project becomes complicated.

- Similarly, use the `receiver.py` file to add all the receiver code in one place for a given Django app.

- Always use the `send_robust` option while dispatching signals unless the use case is specified. Multiple receivers can listen to a given dispatched signal, so it is possible that one of the receivers can fail while executing, and this can cause the next receiver code to not execute. But if we pass `send_robust` while dispatching a signal, then the signal is automatically propagated to the next receiver, even if the previous receiver failed.

- Prevent duplicate signals. The signal dispatch code can be executed multiple times. As it is important to avoid duplicate signals, using `dispatch_uid` can mitigate this issue.

- It can be a nightmare for developers to debug signals in a large code base, so use signals responsibly and always try to add appropriate comments on the use case of signals.

- Avoid the `pre_save` and `post_save` signals if possible. As the project grows complex and multiple developers start to contribute, `pre_save` and `post_save` signals become harder to keep track of. It is always advisable to create a wrapper signal with a similar implementation, which can be easier to trace as the project grows.

We have just learned about a few best practices to be followed while using Django signals. Now, let us learn about Django middleware and how you can create your own custom middleware.

Working with Django middleware

Django middleware is a regular Python class that is hooked into Django's request/response cycle. Middleware can be used to modify a request before Django views process them or modify a response before sending it back to the client. Every middleware is called twice during the request/response cycle – the first time is when the request is received from the client, and then the second time is when the response is sent back to the client.

By default, Django provides a lot of middleware and we also learned, in *Chapter 3*, how DRF also utilizes the Django middleware framework to integrate REST APIs. The default Django middleware is mainly used for authentication and is security-related. When we create a new project, Django automatically plugs in this middleware in the `settings.py` file:

```
MIDDLEWARE = [
    "django.middleware.security.SecurityMiddleware",
    "django.contrib.sessions.middleware.SessionMiddleware",
    "django.middleware.common.CommonMiddleware",
    "django.middleware.csrf.CsrfViewMiddleware",
    "django.contrib.auth.middleware.AuthenticationMiddleware",
    "django.contrib.messages.middleware.MessageMiddleware",
    "django.middleware.clickjacking.XFrameOptionsMiddleware",
]
```

It is important to note that the order in which these middleware items are defined in the `settings.py` file is the same order in which they are evaluated when a request comes and the opposite order when the response is sent back to the client.

Django middleware makes use of the simple Python class concept. The `middleware` class is instantiated when the Django server is initiated, and then for every request, it executes all the code written inside the `__call__` method. Django also provides a way to create our own custom middleware that can be easily plugged into the system. Let us learn how we can create custom middleware in Django.

Creating custom middleware

Custom Django middleware should be used to execute code that we want to run for all the requests/responses. For example, we want to add a custom header to all the requests/responses, or we want to add logging or monitoring for requests. In *Chapter 6*, we learned how to use custom middleware to add get user_id for each request for logging. In this section, we will learn about the core concepts of custom middleware in Django.

Let us now check how to create a custom middleware in Django:

1. Create a file called custom_middleware.py (any other name can also be given). Add the following code to the common/custom_middleware.py file:

```python
class CustomMiddleware:
    def __init__(self, get_response):
        self.get_response = get_response
    def __call__(self, request):
        print("custom middleware before request view")
        # Code to be executed for each request
        response = self.get_response(request)
        # Code to be executed for each response
        print("custom middleware after response view")
        return response
```

In the preceding code snippet, we have two print statements that depict a custom middleware. In our example, we have added a print statement, but we can extend this concept to implement any logic. We can modify the request or response object too using middleware. For example, Django adds the user attribute to the request object using the default authentication middleware.

It is important to note the response = self.get_response(request) code that is responsible for ensuring that we are sending the code execution to the next middleware or the request view logic for execution. The result given by self.get_response is the response returned by the view or the previous middleware.

2. Add the custom middleware to the middleware configuration in the settings.py file. The order in which we are adding the middleware is very important:

```python
MIDDLEWARE = [
    "django.middleware.security.SecurityMiddleware",
    "django.contrib.sessions.middleware.SessionMiddleware",
    "django.middleware.common.CommonMiddleware",
    "django.middleware.csrf.CsrfViewMiddleware",
    "django.contrib.auth.middleware.AuthenticationMiddleware",
    "django.contrib.messages.middleware.MessageMiddleware",
    "django.middleware.clickjacking.XFrameOptionsMiddleware",
```

```
        # Custom middleware
        "common.custom_middleware.CustomMiddleware",
    ]
```

For every request made, our `CustomMiddleware` class will be executed now. To verify the `CustomMiddleware` integration, run the Django server, and then open any URL on the browser (for example, `http://127.0.0.1:8000/blog/paginated/`). We will see the following output in our terminal:

```
Starting development server at http://127.0.0.1:8000/
Quit the server with CONTROL-C.
custom middleware before request view
custom middleware after response view
[03/Dec/2023 12:27:31] "GET /blog/paginated/?page=2&page_size=2
HTTP/1.1" 200 10734
```

The custom middleware prints out every time a request is made to our Django server.

Our custom middleware can be used to implement different logic other than the printing out that we have implemented. Multiple third-party packages use the middleware concept to modify any `request/response` object. Though middleware is a very important concept in Django, you should also remember the performance overhead that middleware brings. A middleware would execute on every request, hence we should understand our use case before writing/using middleware, or else it can degrade our performance.

Summary

In this chapter, we have learned about three advanced concepts in Django. We have learned about the different concepts of pagination and how Django and DRF make use of pagination to improve the RESTful APIs. Next, we learned about the concept of Django signals and how Django provides out-of-the-box signals to decouple different application logic. Django also provides the interface to extend custom signals. Lastly, we learned about the Django middleware concepts. Django middleware is used to modify `request/response` objects. We also learned how to create custom middleware and use it in our Django project.

In this chapter, we have learned all the advanced concepts of Django that any developer should be comfortable implementing, in order to write scalable Django applications.

In the next chapter, we will focus on learning about distributed task queue programming with Django. Celery is one of the most popular Python frameworks that is used along with Django to implement long-running logic. As we build applications, we will come across use cases that are long-running or tasks that need to be scheduled. Celery helps in implementing these long-running/periodic tasks.

8

Using Celery with Django

We have learned about the Django and DRF features that are provided out of the box in our previous chapters. Django and DRF both provide a scope of customization and extendibility to the framework, but async programming has been missing from Django for quite some time. However, with the introduction of Django 4, Django has started to natively support async programming using different async interfaces. This still doesn't help with requirements to execute long-running jobs or jobs executed at periodic intervals. To address this requirement, we use a well-known third-party package in Python, **Celery**.

Celery is a simple, flexible, and reliable distributed system used to process vast amounts of messages. It is a task queue with focus on real-time processing while also supporting task scheduling. In this chapter, we shall learn how to set up Celery with Django and use *celery beat* to perform periodic tasks in Django. Since Celery itself is an altogether different framework, we shall not dive too deep into the framework, rather focusing on the implementation details with a high-level overview to get you started.

In this chapter, we will learn the following topics:

- Asynchronous programming with Django
- Using Celery with Django
- Using *celery beat* with Django

Technical requirements

In this chapter, we will cover some advanced concepts of asynchronous programming, so it is important that the readers are well versed in the following:

- The difference between synchronous and asynchronous programming, and the advantages of the latter over the former.

- The basic concepts of messaging brokers and task queues. Celery is a framework used to process task queues and task scheduling.

- How to set up Redis and use a remote Redis server in local system (we discussed this in *Chapter 6*).

Asynchronous programming in Django

Django 4 introduced the concept of asynchronous programming in its core framework. While support for async views is introduced in Django, the official documentation (`https://docs.djangoproject.com/en/stable/topics/async/#performance`) states that async views would lead to a performance hit.

Hence, the adaption of native asynchronous programming is not yet mature as of Django 4.x. While most common and basic features will not need any async support, Django as a framework is thus lagging behind other frameworks such as Express and Spring. To work around this problem, Django developers use third-party frameworks to implement asynchronous support. Generally, asynchronous programming is helpful when one has a long-running task that can be executed in the background. For example, when we have to send an email, our Django view would make an external API call to an email service and await the response. This whole process can take a while and there is no added value in making client requests wait for the email service; rather, it is always preferred to perform this task in the background. For such async long-running/background tasks, one can use frameworks such as Celery. Celery is a distributed task queue to offload tasks from the main request-response workflow.

> **Note**
>
> One should remember that Celery itself is a framework and has its own capabilities and a complex architecture. The architecture of Celery is beyond the scope of this book. We shall focus only on the integration of Celery with Django in this chapter. To gain more knowledge of the architecture and advanced concepts of Celery, you are advised to go through the Celery official documentation at `https://docs.celeryq.dev/en/stable/index.html`.

In the next section, we will learn how to use Celery with Django and leverage the capabilities of Celery to perform long-running tasks in the background.

Using Celery with Django

Let us try to understand how Celery works under the hood by taking a small real-world example. Imagine you want to perform 10 different tasks and you have 2 people to work on them. Each person can work on only one particular task at a time, and new tasks might arrive at any given time. To solve this problem, the workers would create a list, and every time a new task comes in, they would add the task to that list. Each time a person finishes their task, you would strike off the task from the list and then assign them the next task from the top. If you add more people to work for you, then you can finish your tasks faster. This is what Celery does.

In Celery, there is a task queue that keeps track of all the tasks coming in – task queues are saved using RabbitMQ or Redis. There are Celery workers that pick tasks from the queue and work on them. Django generates these tasks and pushes them to the task queue, from where Celery picks tasks and executes them using Celery workers.

Let us now see how we can integrate Celery into a Django project. We shall use Redis to create and maintain the Celery task queue in our project. In *Figure 8.1*, we have a minimalistic Celery-Django architecture:

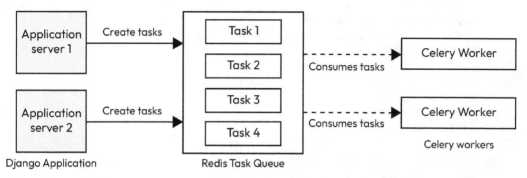

Figure 8.1: Celery architecture with Django

The application servers running in Django create tasks and send those tasks to the Redis task queue. Then the Celery workers pick these tasks one by one from the queue and run them. This is an overly simplified example of how Celery and Django work under the hood, but it should give you a good idea about the architecture.

Let us now see how we can install Celery and integrate it into our system.

Integrating Celery and Django

Celery provides multiple broker integrations such as RabbitMQ, Redis, and Amazon SQS. We shall use Redis as a broker for our Celery service. In *Chapter 6*, we learned how to use a remote Redis server locally, and we will use the same Redis server here for our Celery broker. Let us look into the steps to integrate Celery into Django:

1. As a first step, we need to install Celery in our project using the `pip install celery` command.

2. Create a new `celery.py` file in the same folder where we have the `settings.py` file containing the following code (for us, the location is `backend/config/celery.py`):

```
import os
from celery import Celery

celery_settings_value = "<project name>.settings"
# change <project name> with folder name where your settings.py
file is present.
```

```
os.environ.setdefault("DJANGO_SETTINGS_MODULE", celery_settings_
value)

app = Celery("<project name>")
# change <project name> with folder name where your settings.py
file is present.

app.config_from_object("django.conf:settings",
namespace="CELERY")
app.autodiscover_tasks()
task = app.task

@app.task(bind=True)
def debug_task(self, data):
    print(data)
```

3. Now we have to add the Redis broker connection details to the `settings.py` file. You need to update the Redis connection details in RedisLabs. In this step, we are telling Celery to use Redis as a broker and result backend:

```
REDIS_CONNECTION_STRING = '<user>:<pass>@<endpoint>'
# Use redis lab credentials

CELERY_BROKER_URL = f"redis://{REDIS_CONNECTION_STRING}"
CELERY_RESULT_BACKEND = f"redis://{REDIS_CONNECTION_STRING}"
```

We have integrated Django, Celery, and Redis. Celery uses workers to execute tasks. Now we need to define the workers. Just like the Django management command, Celery also has a management command that can be used to define the workers.

4. Open a new terminal and run the following command (please note – the `<project>` keyword needs to be replaced with the folder name where your `celery.py` file is located):

```
celery --app=<project> worker --loglevel=INFO
```

This command creates a Celery worker that can be used to test and work in the local development environment. If our integration is successful, we should be able to see output shown in *Figure 8.2*:

```
~/djangoinproduction/Django-in-Production/Chapter@8/myblog/backend git:(main) ±54
celery -A config worker --loglevel=INFO

 -------------- celery@Arghyas-MacBook-Pro-2.local v5.3.6 (emerald-rush)
--- ***** -----
-- ******* ---- macOS-14.1.1-arm64-arm-64bit 2023-12-03 15:28:28
- *** --- * ---
- ** ---------- [config]
- ** ---------- .> app:         config:0x1030c2950
- ** ---------- .> transport:   redis://default:**@redis-11193.c281.us-east-1-2.ec2.cloud.redislabs.com:11193//
- ** ---------- .> results:     redis://default:**@redis-11193.c281.us-east-1-2.ec2.cloud.redislabs.com:11193/
- *** --- * --- .> concurrency: 8 (prefork)
-- ******* ---- .> task events: OFF (enable -E to monitor tasks in this worker)
--- ***** -----
 -------------- [queues]
         .> celery           exchange=celery(direct) key=celery


[tasks]
  . config.celery.debug_task

[2023-12-03 15:28:28,575: WARNING/MainProcess] /opt/homebrew/lib/python3.11/site-packages/celery/worker/consumer/consumer.
py:507: CPendingDeprecationWarning: The broker_connection_retry configuration setting will no longer determine
whether broker connection retries are made during startup in Celery 6.0 and above.
If you wish to retain the existing behavior for retrying connections on startup,
you should set broker_connection_retry_on_startup to True.
  warnings.warn(

[2023-12-03 15:28:30,702: INFO/MainProcess] Connected to redis://default:**@redis-11193.c281.us-east-1-2.ec2.cloud.redisla
bs.com:11193//
[2023-12-03 15:28:30,704: WARNING/MainProcess] /opt/homebrew/lib/python3.11/site-packages/celery/worker/consumer/consumer.
py:507: CPendingDeprecationWarning: The broker_connection_retry configuration setting will no longer determine
whether broker connection retries are made during startup in Celery 6.0 and above.
If you wish to retain the existing behavior for retrying connections on startup,
you should set broker_connection_retry_on_startup to True.
  warnings.warn(

[2023-12-03 15:28:32,218: INFO/MainProcess] mingle: searching for neighbors
[2023-12-03 15:28:36,030: INFO/MainProcess] mingle: all alone
[2023-12-03 15:28:41,290: INFO/MainProcess] celery@Arghyas-MacBook-Pro-2.local ready.
```

Figure 8.2: Terminal running Celery workers

Once we have confirmed the Celery worker is running, we can stop the worker process and move to the Django integration step.

5.   To test our integration, let us call our Celery task from a Django view. Add the following code to the blog/views.py file:

```python
from config.celery import debug_task
@api_view(['GET'])
def verify_blog(request):
    verify_word = request.GET.get('verify_word')
    debug_task.delay(f"Celery Task verification: {verify_word}")
    return Response({'status': 'success'})
```

Please note that we are importing debug_task *from* config.celery *because we have created the* celery.py *file inside the* config *folder/app. Your version might have any other name, so make sure to check.*

Now we need to link our view to the URLs. Add the following code to the blog/urls.py file:

```python
urlpatterns = [
    path('verify/', views.verify_blog)
]
```

We have successfully integrated Celery into our Django project.

Now to verify whether our Celery integration is working as expected, let us run the Celery worker on a terminal by running the following command:

```
celery --app=config worker --loglevel=INFO
```

We should see output similar to that shown in *Figure 8.2*. Now let the process run and open a new terminal, where we need to run the `python manage.py runserver` command. Open the `/verify` endpoint in the browser by going to `http://127.0.0.1:8000/blog/verify/?verify_word=abc`. We should get a success response on the browser, and when we go to the Celery worker terminal, we should see output reading `Celery Task verification: abc`, similar to what is shown in *Figure 8.3*:

```
[2023-12-03 15:37:31,441: INFO/MainProcess] mingle: searching for neighbors
[2023-12-03 15:37:35,428: INFO/MainProcess] mingle: all alone
[2023-12-03 15:37:40,348: INFO/MainProcess] celery@Arghyas-MacBook-Pro-2.local ready.
[2023-12-03 15:40:39,488: INFO/MainProcess] Task config.celery.debug_task[4e67553e-48d2-41fc-af60-34fb0a8e6cac] received
[2023-12-03 15:40:39,490: WARNING/ForkPoolWorker-8] Celery Task verification: abc
[2023-12-03 15:40:40,544: INFO/ForkPoolWorker-8] Task config.celery.debug_task[4e67553e-48d2-41fc-af60-34fb0a8e6cac] succe
eded in 1.05518991602002264s: None
[2023-12-03 15:40:43,424: INFO/MainProcess] Task config.celery.debug_task[08976c33-8b35-40c6-b307-f494d8fdb8d8] received
[2023-12-03 15:40:43,431: WARNING/ForkPoolWorker-8] Celery Task verification: abc
[2023-12-03 15:40:43,923: INFO/ForkPoolWorker-8] Task config.celery.debug_task[08976c33-8b35-40c6-b307-f494d8fdb8d8] succe
eded in 0.49289650001446716s: None
```

Figure 8.3: Output from a Celery task that was initiated in Django

> **Important note**
>
> Celery doesn't support hot reloading out of the box. It is common for developers to face a problem where they have updated the code, but the task is not working as expected in the local development server. Please remember that when we start the Celery worker, it loads the code in memory and keeps running it unless we manually restart the Celery worker.
>
> To solve this problem, we can use third-party packages such as Watchdog and create custom commands. Read more about Watchdog at `https://github.com/gorakhargosh/watchdog`.

We have integrated Celery with Django. Let us now learn how we can effectively use Celery in a Django project and examine the different interfaces Celery provides.

Interfaces of Celery

Celery is useful to execute any long-running tasks that Django applications need to run. Whenever Django encounters any long-running/background job to be executed, it lets Celery work on that task instead.

As a good practice, it is always recommended to create separate `tasks.py` files in each app. For example, our blog app should have a `blog/tasks.py` file that contains all the Celery tasks associated with the blog app. Let us consider one use case – whenever a blog gets published, we want to send an email to all the blog's followers. This is an ideal use case where we should let the Celery worker handle the task of sending emails to all the followers. We need to create a `blog/tasks.py` file with the following code:

```
from blog import private
from config.celery import task
@task(bind=True)
def send_email_to_followers(self, author_id, blog_id):
    print('Sending email to all follower logic')
```

In this snippet, we are creating a Celery task called `send_email_to_followers` that sends emails to all the followers of an author whenever they write a new blog.

Now we integrate this Celery task into the Django view. We are calling the Celery task created using the `send_email_to_followers` function and sending it the `author_id` and `blog_id` values as needed:

```
From blog import tasks
def publish_blog(request):
    blog = save_blog(request)
    # publish blog code
    tasks.send_email_to_followers.delay(blog.author_id, blog.id)
```

Please note we are using the `.delay` keyword to call the Celery task. If you don't use `.delay`, then `publish_blog` will behave like a simple Python function and not be executed by the Celery worker.

We have learned how we can create Celery tasks and use them in Django. In our example, we have used the most basic configuration and task definition possible. As your code grows in size and complexity, you may need to integrate more advanced features of Celery.

> **Important note**
> Celery itself is a framework and has a lot of built-in features. We are only focusing on the high-level interface and important points that one should be aware of as an introduction. It is highly recommended that you go through the official documentation and learn about all the advanced features Celery provides. You can find the official documentation at `https://docs.celeryq.dev/en/stable/userguide/index.html`

The advanced features of Celery are out of scope of this book. However, we shall next review a few best practices and advanced concepts that are useful to run Celery and Django in production.

Best practices for using Celery

Let us learn a few best practices that can help us write Celery tasks in production that is scalable. (as stated, we will focus on introducing the concepts and not diving too deep into the details):

- Using `.delay` to call tasks is a good way to get started with Celery. But as we start moving up to the more advanced concepts of Celery, we should start using `.apply_async` to call Celery tasks. Since `.delay` is an additional wrapper over `.apply_async`, a lot of options are abstracted out that can be helpful.

- Use the `autoretry_for` option in the task definition. This is particularly helpful to let Celery retry whenever it encounters any failure.

- Use the `max_retries` option in the task definition to configure the maximum number of retries Celery can perform before marking the task as failed.

- Always use `time_limit` on long-running tasks. This is a hard limit on tasks. After this threshold is crossed, Celery will kill the task automatically. One should implement a hard limit on all tasks so that long-running tasks don't use all the workers and choke the task queue.

- Always have short tasks – split long tasks into smaller chunks of tasks. For example, if you have to send emails to 100 users, create a task that would again spawn multiple tasks to send the individual emails to each user.

- Creating Celery tasks that guarantee idempotency and atomic behavior is important to create a scalable Celery system.

- Always make sure to create backward-compatible Celery tasks. For example, if a Celery task takes three arguments in the task definition, such as `def add_task(a,b,c)`, and now we want to remove the third argument, then all the existing tasks created with the three arguments will start failing. Hence, we should make sure to write backward-compatible code so that existing tasks won't break with future changes.

- As your project grows, you might start using Celery heavily. Celery uses a queue data structure to store the tasks and executes them using **first in, first out** (**FIFO**) order, but there might be use cases where you need to prioritize certain tasks over others. This can be performed by using a different queue, and also by setting the priority while creating tasks. Please read the official documentation on priority for further details at `https://docs.celeryq.dev/en/stable/userguide/routing.html#redis-message-priorities`

- Django signals are synchronous in nature, but by using Celery we can create asynchronous signals.

- Do not pass database objects to tasks, since a database object can be updated by another process. To avoid working on stale data, always pass the object ID to the task. This ensures that the task always gets the latest data to perform operations on.

- Celery assigns task names by default using the module and task name. While this works perfectly when you have good naming conventions, it can cause issues in debugging if the task names are not added properly. Celery has a `name` interface to let you provide custom names for Celery tasks if required.

- If you have important business logic dependent on Celery, always have reconciliation jobs that run periodically to make sure Celery tasks have executed as expected.

We have now learned how to integrate Celery into Django projects and the best practices to follow while implementing Celery in production. There is one more important use case that Celery helps developers tackle – periodic tasks. *celery beat* is a scheduler that checks at regular intervals for any periodic task and creates Celery tasks for the workers to execute.

Using celery beat with Django

Periodic tasks are one of the most common use cases for Celery. For example, say you want to send periodic emails to your users every week, or create certain summary reports every week. For such repetitive tasks, Celery provides **celery beat**, which is a scheduler that keeps checking for any scheduled task to be executed at the given time and then runs it. The *celery beat* worker would create tasks in place of the Django application and put them in the Redis task queue, from where tasks are then picked up by the Celery workers and executed.

Let us learn how to integrate *celery beat* into our Django project:

1. There are multiple ways to create periodic tasks, such as adding a `crontab` configuration in the settings file, or plugging a beat schedule into the `app.conf.beat_schedule` object. They work perfectly, but lack ease of use. One option that truly stands out is using the `on_after_finalize.connect` signal.

   In the following example, we create a periodic task that runs once every minute and triggers `demo_task`. Add the following code to the `author/tasks.py` file:

```
from config.celery import app
from config.celery import task
from celery.schedules import crontab
@app.on_after_finalize.connect
def setup_basic_periodic_tasks(sender, **kwargs):
    sender.add_periodic_task(
        crontab(),
        demo_task.s("periodic task")
    )


@task(bind=True)
def demo_task(self, value):
    print(value)
```

We can change the interval by using `crontab` syntax. For more details, check the official documentation of *Celery crontab* at `https://docs.celeryq.dev/en/main/userguide/periodic-tasks.html#crontab-schedules`

2. *celery beat* is a separate process that continuously monitors for any scheduled task. We need to run a new Celery command on our terminal to run the *celery beat* service. Please note that the `<project>` keyword needs to be replaced by the folder name in which the `celery.py` file is present (in our case it is `config`), just like we did for our Celery setup:

```
celery --app=<project> beat --loglevel=INFO
```

Now we have a *celery beat* instance running that will keep checking for any periodic tasks and then push those tasks to the Redis queue, from which the Celery worker will pick tasks and execute them. Thus, we have integrated *celery beat* into our Django project to run periodic tasks.

Let us verify our *celery beat* integration project. Since we have already created a periodic task that runs every minute, we should be able to check for output in the terminal. First, start the *celery beat* server by running the following command in the terminal:

```
celery --app=config beat --loglevel=INFO
```

Let *celery beat* run in a terminal, and now open a new terminal and start the Celery worker by running the following command:

```
celery --app=config worker --loglevel=INFO
```

Now we have two Celery instances running: one is running the *celery beat* service and the other one is running the Celery worker. We should be able to see the output of our *celery beat*, similar to what is shown in *Figure 8.4*, with a `demo_task` instance created every minute:

```
~/djangoinproduction/Django-in-Production/Chapter08/myblog/backend git:(main) ±57
celery --app=config beat --loglevel=INFO

celery beat v5.3.6 (emerald-rush) is starting.

LocalTime -> 2023-12-03 17:30:16
Configuration ->
    . broker -> redis://default:**@redis-11193.c281.us-east-1-2.ec2.cloud.redislabs.com:11193//
    . loader -> celery.loaders.app.AppLoader
    . scheduler -> celery.beat.PersistentScheduler
    . db -> celerybeat-schedule
    . logfile -> [stderr]@%INFO
    . maxinterval -> 5.00 minutes (300s)
[2023-12-03 17:30:16,602: INFO/MainProcess] beat: Starting...
[2023-12-03 17:30:17,368: INFO/MainProcess] Scheduler: Sending due task author.tasks.demo_task('periodic task') (author.tasks.demo_task)
[2023-12-03 17:31:00,010: INFO/MainProcess] Scheduler: Sending due task author.tasks.demo_task('periodic task') (author.tasks.demo_task)
[2023-12-03 17:32:00,000: INFO/MainProcess] Scheduler: Sending due task author.tasks.demo_task('periodic task') (author.tasks.demo_task)
[2023-12-03 17:33:00,000: INFO/MainProcess] Scheduler: Sending due task author.tasks.demo_task('periodic task') (author.tasks.demo_task)
[2023-12-03 17:34:00,000: INFO/MainProcess] Scheduler: Sending due task author.tasks.demo_task('periodic task') (author.tasks.demo_task)
```

Figure 8.4: Logs showing celery beat creating demo_task every minute

celery beat is responsible for creating tasks, but the Celery worker is responsible for the execution of tasks. We can check the output of the Celery worker showing `demo_task` executed every minute as shown in *Figure 8.5*:

```
~/djangoinproduction/Django-in-Production/Chapter08/myblog/backend git:(main) ±57
celery --app=config worker --loglevel=INFO
[2023-12-03 17:30:33,222: INFO/MainProcess] celery@Arghyas-MacBook-Pro-2.local ready.
[2023-12-03 17:30:33,976: INFO/MainProcess] Task author.tasks.demo_task[2019ef62-33d1-4986-8e1b-9e3170a5d919] received
[2023-12-03 17:30:33,981: WARNING/ForkPoolWorker-8] periodic task
[2023-12-03 17:30:35,058: INFO/ForkPoolWorker-8] Task author.tasks.demo_task[2019ef62-33d1-4986-8e1b-9e3170a5d919] succeeded in 1.0782499169872608s: None
[2023-12-03 17:31:01,050: INFO/MainProcess] Task author.tasks.demo_task[aeee9f59-0d65-499c-9ce7-abf12f464c4d] received
[2023-12-03 17:31:01,053: WARNING/ForkPoolWorker-8] periodic task
[2023-12-03 17:31:01,557: INFO/ForkPoolWorker-8] Task author.tasks.demo_task[aeee9f59-0d65-499c-9ce7-abf12f464c4d] succeeded in 0.5043243749823887s: None
[2023-12-03 17:32:00,761: INFO/MainProcess] Task author.tasks.demo_task[06bc8964-4e81-4df9-b151-25ff019eb34b] received
[2023-12-03 17:32:00,763: WARNING/ForkPoolWorker-8] periodic task
[2023-12-03 17:32:01,267: INFO/ForkPoolWorker-8] Task author.tasks.demo_task[06bc8964-4e81-4df9-b151-25ff019eb34b] succeeded in 0.5043796249956358s: None
[2023-12-03 17:33:00,985: INFO/MainProcess] Task author.tasks.demo_task[d9671f86-cfe3-4c7b-ac85-b59a6fb107cf] received
[2023-12-03 17:33:00,988: WARNING/ForkPoolWorker-8] periodic task
[2023-12-03 17:33:01,494: INFO/ForkPoolWorker-8] Task author.tasks.demo_task[d9671f86-cfe3-4c7b-ac85-b59a6fb107cf] succeeded in 0.5059737499977928s: None
[2023-12-03 17:34:00,738: INFO/MainProcess] Task author.tasks.demo_task[3496f8cd-7957-4581-9b66-a512358274ee] received
[2023-12-03 17:34:00,991: WARNING/ForkPoolWorker-8] periodic task
[2023-12-03 17:34:01,497: INFO/ForkPoolWorker-8] Task author.tasks.demo_task[3496f8cd-7957-4581-9b66-a512358274ee] succeeded in 0.5062688339967281s: None
[2023-12-03 17:35:01,092: INFO/MainProcess] Task author.tasks.demo_task[b37903c5-10d2-4f52-999c-fed3791bc598] received
[2023-12-03 17:35:01,097: WARNING/ForkPoolWorker-8] periodic task
[2023-12-03 17:35:01,602: INFO/ForkPoolWorker-8] Task author.tasks.demo_task[b37903c5-10d2-4f52-999c-fed3791bc598] succeeded in 0.5054065829899628s: None
```

Figure 8.5: Logs showing the Celery worker executing a demo task

> **Important note**
>
> One should always run a single instance of *celery beat* at any given time. If we ran multiple instances of *celery beat*, we would see duplicate execution of periodic tasks.
>
> However, we might want to run backup instances of *celery beat* to protect against any failure. For such scenarios, we can use a third-party package called `celery-redbeat`. For further details, please review the documentation provided by the `celery-redbeat` package at `https://redbeat.readthedocs.io/en/latest/index.html`

celery beat is one of the most powerful tools with which to automate repetitive tasks. We have seen how we can integrate Django with Celery and *celery beat*. In *Chapter 11*, we shall learn more about how we can Dockerize the whole Django, Celery, and *celery beat* setup to improve the overall developer experience. Also, it is important to monitor Celery workers and the *celery beat* service. We should also perform regular health checks on both the Celery worker and *celery beat*. We shall discuss these in *Chapter 14*.

Summary

In this chapter, we learned how to tackle the problem of long-running tasks using Celery. In web development, having a long-running task is quite common, such as when we have to generate a monthly report, send an email to a batch of users, process data from an uploaded file, and so on. Celery is very handy to address such use cases. Similarly, there are plenty of use cases where you need task-scheduling functionality. For example, if we wanted to send a weekly consolidated report to users every week, send reminder emails to end users, and so on, *celery beat* would be our go-to tool to build such features.

In *Chapter 9*, we will learn how to write unit and integration tests with Django. Writing unit tests is one of the most important yet frequently overlooked phase in the development cycle. We will learn all the best practices to follow while writing unit and integration tests.

9

Writing Tests in Django

In *Chapter 8*, we learned about Celery and how we can write Celery tasks to solve our problem of long-running tasks. Celery helps us to implement periodic tasks using Celery Beat. So, with all the knowledge we have gained so far, we can create applications that solve business use cases. One thing we should always remember is that writing code is easy, especially when doing so from scratch, but the hard part comes when we have to maintain the code and add new features/improve features in the existing code base. This is where one of the most important concepts of programming enters the picture – **testing**.

In this chapter, we shall learn the different aspects of testing and why testing is important, learning how to write maintainable test cases that don't make your system brittle but give you the flexibility and confidence to ship new features fast.

Here are the topics we'll be covering in this chapter:

- Introducing the different types of tests in software development
- Setting up tests for Django and DRF
- Learning the best practices to write tests in Django Project

Technical requirements

In this chapter, we shall learn about the concepts of testing; we expect you to be well-versed in the concepts of Django, as covered in the previous chapters. We assume that you to have a basic understanding of testing in the software development domain. Also, you should have a basic understanding of how to write test cases in Python and have some experience in writing unit/integration test cases for your code.

Here is the GitHub repository that has all the code and instructions for this chapter: `https://github.com/PacktPublishing/Django-in-Production/tree/main/Chapter09`.

> **Important note**
>
> In this chapter, we are learning test cases, and we will be writing/taking code examples from the previous chapter. Since we are covering a lot of concepts, it is not possible to include all the code snippets in this chapter. To follow along better, it is recommended that you follow the code sample saved in the GitHub repo. Also, feel free to reach out to me via Discord for any queries: `https://discord.gg/r7UDTKbgQw`.

Introducing the different types of tests in software development

All of us developers love to write code and build systems and applications. All of us hate attending meetings. Another thing most developers hate is *writing tests*. We all have been there. When I started my career, and whenever I was asked by my senior/mentor to write test cases for code, I would dread the thought. It would take almost double the time to write test cases for the code than it took to build the feature.

But *tests are a necessary evil*, and now that I am a senior engineer building dozens of scalable systems, I always prefer to write tests for any project I work on. This gives me the confidence to ship my code to production without the fear of breaking the existing system, and this helps in the faster shipping of features.

When we build new features, we always test the code manually. For example, if I had to write code to find out whether a number is prime or not, I would write the code and then pass different values to it to test whether it works as expected or not. Then, let's say that the requirement got updated – I would now have to find out whether the input is prime and greater than 100. Again, I would have to update the code and manually test my logic by entering different inputs. Then, the requirement got updated again – I would then have to find a prime number that is greater than 100 but fewer than 200.

This is not an uncommon use case in software development. As we start building systems, we are expected to receive updated requirements, and we have to make sure our code works for the previous logic as well as for the new logic. We just saw the example of a prime number finder. Imagine if we had to do a similar thing for a major application such as a flight booking system. It would be a nightmare to test all the corner cases manually. That is where the concept of "automated testing" comes into the picture. By writing tests, we can make sure that our code executes in all the use cases and that our systems are stable enough.

The advantage of writing unit tests comes to the limelight once a project becomes mature. I have experienced it first-hand while working on multiple projects. In the initial few months, feature development and testing almost have a 1:1 ratio – for example, if a developer spends one hour writing code for the feature, they might spend the same hour writing test cases for the code. As the project grows, this remains consistent, and then the main question you would ask is, why spend 2x time when the team can perform the same task in half the time? The answer lies with the team that doesn't write test cases for their code. For the initial couple of months, the team can write the same feature in half the time, but as the project grows, the time to deliver the feature increases exponentially, since they have to manually test all the use cases and also introduce more bugs in the production that would need more production code revert and more time to debug the issues. So, a team writing test cases is a win in the long run, with faster delivery and a more stable system.

There are multiple types of testing in software development, such as **unit testing**, **integration testing**, **regression testing**, **security testing**, and **user acceptance testing**. We shall only discuss the major types of automation testing that a developer should be responsible for.

Unit testing

Unit testing is a method of testing where we write tests for the smallest units of our code. Unit tests should be written by developers and are stored in the same repository as the source code. These are automated tests that run whenever there is any update to code. For example, if we have a blog application, we write multiple unit tests for our blog feature – we write a unit test for the function that calculates the word count of our blog. Unit tests are the smallest unit of testing – each method/function should do just one task that can be easily tested by a written unit test. If you can write unit tests easily, it that means your code is well-written and structured well in smaller modules. It is often recommended to rewrite your application code as smaller decoupled modules if you are unable to write unit tests properly, since a well-defined module is always easy to write unit tests. We shall learn in detail how to write unit tests for Django in this chapter.

Integration testing

Integration testing is a type of testing where we test how the different units of our system work together. It involves testing the functionality of a system end to end. For example, our integration test would make sure we can save the blog along with the process data in our database when we pass raw content to a function or API. Integration testing is a vast topic, and you should write as many integration tests as possible. Deciding on what functionalities developers should cover in the integration test comes with experience.

E2E testing

End-to-end (**E2E**) testing is a type of testing where we verify the working of a feature from start to finish. Writing E2E test cases helps us ensure our business logic works correctly and systems function properly. We won't deep-dive into how to write E2E tests, but we'll let you explore different techniques to write them. You can write an E2E test using Postman to write Postman scripts that would verify a contract (`https://learning.postman.com/docs/writing-scripts/test-scripts/`). You can also use other third-party tools, such as Playwright (`https://playwright.dev/`) or Puppeteer (`https://pptr.dev/`), to write E2E tests.

Now, let us deep-dive into how we can set up tests in our Django project and learn how to write basic tests for our modules.

Setting up tests for Django and DRF

In this chapter, we shall primarily focus on how to write unit tests and integration tests for Django and DRF projects. In the Python and Django community, there is a long-standing debate over which of the following packages is better for writing tests for projects:

- `unittest` (https://docs.djangoproject.com/en/stable/topics/testing/) comes shipped with Django and is also an extension of Python's `unittest` package

- `pytest-django` (https://pytest-django.readthedocs.io/en/latest/) uses `pytest` under the hood and provides a lot of abstraction, which takes care of all the basic nuances of writing tests

We shall not get into the debate and allow developers to decide which package is better from their own experience and preference. We shall take the simple route and follow what the Django official documentation recommends, using the default Django `unittest` to set up tests and learn them.

Let us now get started with writing test cases in Django. Django and DRF both use Python's standard testing module, `unittest`. Throughout this book, we have primarily focused on writing APIs for our Django application using DRF; hence, we shall focus on writing test cases for APIs written in DRF in this section. DRF extends the default Django test case classes for APIs and provides us with an interface to write test cases. The official documentation for DRF mentions all the available test case classes (https://www.django-rest-framework.org/api-guide/testing/). Let us perform the initial setup and configuration.

Structuring and naming our test cases

When we create a Django app in our Django project, Django by default creates a `tests.py` file inside the app folder. We shall not use the `tests.py` file to write our test cases, since it would be difficult to write all the test logic for one application in one single file. Rather, we shall delete this file and create a new package, named `tests`, inside our Django app. Inside this `tests` package, we shall create new files that contain all the test cases, and these files should start with the `test` keyword, such as `test_api.py` or `test_public.py`, so that the Django `unittest` framework can detect the test files automatically. For example, in *Figure 9.1*, we can see how to place the `tests` folder inside the `author` app and then create multiple `test_<>.py` files that contain all the corresponding test code. We have the `api.py` tests written in `test_api.py`, `public.py` tests written in `test_public.py`, and so on.

Figure 9.1: The file structure for tests in the Django app

> **Important note**
>
> We should always make sure our test files start with `test`; otherwise, the Django `unittest` framework will not be able to detect the test cases automatically. Often, developers make the mistake of *appending* `test` rather than *prepending* `test`. This way, developers assume they have written tests and they have passed, but in reality, these test files are never detected by Django, and their code remains without a test case. In the later part of the chapter, we discuss how we can write automation scripts to catch these issues.

Test cases will grow as a project matures, and it is normal to have test cases 3x–4x the scale of the original code; hence, we should also break test files further. We can create a package and follow the same file naming convention of starting with `test_<api>.py`, as shown in *Figure 9.2*.

Figure 9.2: An example of a test package, test_api

Django's `unittest` will detect all tests inside subfolders and run the test files that start with `test`. Now that we have learned how to create test files and name them appropriately, let us explore setting up a database to write test cases.

Setting up a database for tests

While writing most of our test cases for Django applications, we would have to use a database. It might be tempting to use SQLite or a similar database to use to achieve a faster run of test cases. However, you should avoid performing any such deviation from production because we write test cases to catch any issues early in our development process. By using SQLite or a similar database, we likely achieve faster test execution, but our test environment will be nowhere similar to what our production environment is, and we might face data integrity issues on production, which won't get caught by our test cases.

For all the tests that need a database, Django creates a separate database and does not use the "real" database that is used for regular requests. The default test database created by Django will have `test_` prepended to the `NAME` value, mentioned in the `DATABASES` configuration of the `settings.py` file.

However, if we want to add a specific name to our test database, then we can do so by adding the `NAME` config inside the `TEST` attribute:

```
DATABASES = {
    "default": {
        "ENGINE": "django.db.backends.postgresql",
        "USER": "mydatabaseuser",
        "NAME": "mydatabase",
        "TEST": {
            "NAME": "mytestdatabase",
        },
    },
}
```

It is always advisable to be verbose when naming a test database so that engineers can easily find the database name. By default, after running tests every time, Django will destroy the database created during the execution of the tests; this will happen regardless of whether the tests pass or fail. If we want to persist the data from the previous test case, we should run the following command:

```
python manage.py test --keepdb
```

This is not an ideal approach, and developers should be cautious while adding a `keepdb` flag, since data from previous tests are now present in the database and might cause some unpredictable behavior.

> **Important note**
>
> Note Django will need to create and delete a test database every time it runs test cases. The Postgres connection details used in the `settings.py` file should have permission to create and delete databases. If you are using ElephantSQL, the free tier doesn't allow you to create multiple databases, and you would face an error while running the test cases. To fix this, you can use a local Postgres setup or other online services, such as Neon (`https://neon.tech`). For more details on using Neon, please check the instructions at `https://github.com/PacktPublishing/Django-in-Production/tree/main/Chapter09#setting-up-a-database-for-tests`.

Now that we have set up the database and learned where to create test cases, let us create a basic test case in DRF.

Writing basic tests in DRF

Let us write a basic test for an API that sends a static response. First, let us have a basic API that returns `"hello world!"` as a response to a GET request. Add the following code to the `blog/views.py` file:

```
from rest_framework import status
from rest_framework.decorators import api_view
from rest_framework.response import Response


@api_view(['GET'])
def basic_req(request):
  if request.method == 'GET':
    resp = {"msg": "hello world!"}
    return Response(data=resp, status=status.HTTP_200_OK)
```

The URL is configured as `/blog/hello-world/`, by adding the following code to the `blog/urls.py` file:

```
urlpatterns = [
    path("hello-world/", views.basic_req),
]
```

We have now created a basic URL that responds with a `"hello world"` response. We will now write a test case to verify the response for this API. In the following example, we shall use DRF's default testing framework to write tests for our API:

```
from django.urls import reverse
from rest_framework.test import APITestCase
from rest_framework import status
```

```python
class BasicTests(APITestCase):
  def test_basic_req(self):
    """
    Test the basic url path response.
    """
    #ARRANGE-Create a URL and expected response.
    url = '/blog/hello-world/'
    expected_data = {"msg": "hello world!"}

    # ACT - Perform API call by DRF's test APIClient.
    response = self.client.get(url, format='json')
    # ASSERT- Verify the response.
    self.assertEqual(response.status_code, status.HTTP_200_OK)
    self.assertEqual(response.data, expected_data)
```

Once we have written the preceding code, we can execute the test by using Django's management command. Django will identify automatically the test files and execute them:

```
python manage.py test
```

Running the preceding command will execute the tests and show the test case results.

We are not getting into the basics of how test cases run and what a success or failure output looks like, since we are expecting you to have an understanding of Python test cases.

In our example, we have followed one of the most commonly used testing patterns, **Arrange-Act-Assert**:

1. **Arrange** the input and anything that is needed to set up the test cases. We have defined the URL and the expected response data in our *arrange* step. Any data or setup that is needed specifically for the test should be added at the top of the test case. We first set up the URL and define the `expected_data` value.

2. **Act** on the target behavior. In this section, we make the appropriate call to the actual application code and capture the result in each variable. For our case, we make an API call to the `hello-world` path using DRF's `APIClient` test interface. If we are writing a unit test, then in the *act* step, we call the function and save the returned value inside a variable.

3. **Assert** the expected data. In our *act* step, we captured the data from the target URL. Now, in this step, we verify whether our response value is the same as what we expect. For example, the status code expected in our API is `200`, and the response data message should be `hello world!`; we verify the API status code and response data in this step.

DRF's official documentation has more examples of writing test cases for different APIs and use cases. Read more at `https://www.django-rest-framework.org/api-guide/testing`.

Let us now focus on the advanced implementation of testing.

Writing tests for advanced use cases

Application code is never just a basic API that sends us a response; in real life, our application will have all types of complex logic, and we will write code for a lot of advanced topics. In this section, we shall learn how to write test cases for advanced use cases such as authentication, models, Celery, and signals.

Testing API authentication

Most of the time, the APIs we create would require some kind of authentication. When we write tests for our API, we need to also make sure our API tests for authentication and authorization. DRF's `APIClient` provides different interfaces to test authentication:

- `self.client.login(username=<string>, password=<string>)`: This allows us to authenticate the test client request as a regular login request that uses `SessionAuthentication`.

- `self.client.credentials(HTTP_AUTHORIZATION=<header string>)`: This allows us to authenticate the `APIClient` test request using HTTP headers. We discussed DRF API token-based authentication in *Chapter 5*. Now, we will look at an example to see how we can write a login using an API token for our `APIClient` test. In the following example, we have `test_unauthenticated_req`, which will show us what happens when the request is not logged in. We are expecting the response code to be `401`. In `test_authenticated_req`, we log in to the request by using `self.client.credentials`. This helps us to assert that the API will work only when authenticated by the right token in the header:

```python
class BasicTests2(APITestCase):
  def test_unauthenticated_req(self):
    url = '/blog/hello-world-2/'
    response = self.client.get(url, format='json')

    # Since user is not logged in it would get 401.
    self.assertEqual(response.status_code, status.HTTP_401_
UNAUTHORIZED)

  def test_authenticated_req(self):
    url = '/blog/hello-world-2/'
    expected_data = {"msg": "hello world!"}
    user = User.objects.create_user( username='demouser',
password='demopass')
    token, created = Token.objects.get_or_create(user
='demouser')

    # Login the request using the HTTP header token.
    self.client.credentials(HTTP_AUTHORIZATION= f'Token {token.
key}')
    response = self.client.get(url, format='json')
```

```
    # User is logged in we would get the expected 200 response
code.
    self.assertEqual(response.status_code, status.HTTP_200_OK)
    self.assertEqual(response.data, expected_data)

def test_wrong_authenticated_req(self):
    url = '/blog/hello-world-2/'

    # Login to the request using a random wrong token.
    self.client.credentials(HTTP_AUTHORIZATION= f'Token random')
    response = self.client.get(url, format='json')

    # Request has wrong token so it would get 401.
    self.assertEqual(response.status_code, status.HTTP_401_
UNAUTHORIZED)
```

It is important to add test failure and success to the token verification. In `test_wrong_authenticated_req`, we verify that our request checks the proper token for authentication. By adding a random token, we get an unauthenticated response. This asserts that our authentication system works for the given API.

- `self.client.force_authenticate(user=<obj>, token=<obj>)`: When we write tests for APIs with complex responses, we might not want to test the authentication flow but, instead, forcefully authenticate the request. For such use cases, DRF provides the `force_authenticate` interface. If we pass a `user` object to the `force_authenticate` method, then the `user` object is automatically assigned to the `request.user` attribute. In the following example, we use the `force_authenticate` method to log the `a1` user into the request, and the `request.user` object will have an `a1` user object:

```
def test_force_authenticate_with_user(self):
    """
    Setting `.force_authenticate()` with a user forcibly
authenticates.
    """
    u1 = User.objects.create_user('a1', 'a1@abc.co')
    url = '/blog/hello-world-2/'

    # Forcefully login and update request.user
    self.client.force_authenticate(user=u1)
    response = self.client.get(url)

    # Tests work with the login user.
    self.assertEqual(response.status_code, status.HTTP_200_OK)
    self.assertEqual(response.data, expected_data)
```

Similarly, we can pass a `token` object to update the corresponding `request.user`, or we can pass both the `user` and `token` objects to update the `request` object attribute.

We have seen three different methods provided by DRF to test the authentication workflow and make our test requests as logged-in requests. Always remember to write at least one test that verifies the whole authentication workflow, and use the `force_authenticate` request to mark the requests as logged in.

When writing tests for different scenarios, we use the `force_authenticate` method each time we write a logged API; this means we have to repeat the same code of forceful authentication again and again. To solve this problem, we should use the `setUp` method that comes from the base class of `unittest.Testcase`.

Setting up and tearing down test cases

Whenever we write test cases for a particular API, it may be possible that we have to perform the same set of tasks to set up the environment – for example, creating the data or logging in a user. Such tasks should be implemented in the `setUp` method. In the following example, we perform a forceful login to the client request in `setUp`, so every time a new test runs from the `BasicTests` class, `setUp` will run and log into the request; this helps us to make a login request every time we write a test. Similarly, if we want to perform an operation every time after we execute the test, we can use the `tearDown` method. For example, if we want to log out a user after every test request run, we use this method:

```python
class BasicTests3(APITestCase):

    def setUp(self):
        self.url = '/blog/hello-world/'
        user = User.objects.create_user('al', 'al@abc.co')
        self.client.force_authenticate(user=user)
        print('Running Setup')

    def test_with_setup_authenticated_req(self):
        print('test 1 running')
        expected_data = {"msg": "hello world!"}
        response = self.client.get(self.url, format='json')

        # User is logged in, expected 200 response code.
        self.assertEqual(response.status_code, status.HTTP_200_OK)
        self.assertEqual(response.data, expected_data)

    # demo test
    def test_demo(self):
        print('test 2 running')
        self.assertEqual(1,1)
```

```
def tearDown(self):
  self.client.logout()
  print('Running teardown')
```

In the preceding example, we have added `print` statements to understand the order of execution in our test cases. We can see that `setUp` and `tearDown` execute twice, since we have two test cases.

As we write test cases, we will have to interact with different model objects to write better test cases, and it becomes tricky to create model objects that have foreign key relations with other models. Also, developers often face problems creating the initial setup of data in a database that can help them write tests for the business logic. To solve this problem, developers might want to use fixtures; however, at scale, fixtures are difficult to manage, so I would recommend using the following package as a substitute for fixtures, `factory_boy` (`https://github.com/FactoryBoy/factory_boy`).

Using factory_boy

`factory_boy` is a fixture replacement that can easily create complex DB objects in Django. Developers don't now have to manually create model objects; they can focus on writing core test cases, allowing `factory_boy` to create all the required model dependencies under the hood. The `factory_boy` package is a generalized module in Python that can work with different Python-based ORMs. We shall primarily focus on creating model objects using the Django ORM.

Let us look at an example of how we can create a model object that has a foreign key relationship with another model. We will define two model factories, `AuthorFactory` and `BlogFactory`. They are linked directly to the Django model objects defined in the `Meta` class property. Do remember to inherit `DjangoModelFactory` for each class definition. We shall save the code in `factoryboy.py` or any other relevant filename, but always make sure you add all the factory boy definition code in a separate file so that you can clearly bifurcate the factory code from other test files. This separation of files will also enable us to avoid any circular dependency issues, since we might use `SubFactory` to call other `factoryboy.py` files from different apps:

```
from factory import SubFactory, Sequence
from factory.django import DjangoModelFactory


from blog import models as blog_models
from author import models as author_models


class AuthorFactory(DjangoModelFactory):
    class Meta:
        model = author_models.Author
    name = Sequence(lambda n: f'Author {n}')
    email = Sequence(lambda n: f'a{n}@gm.com')
class CoverImageFactory(DjangoModelFactory):
    class Meta:
```

```
        model = blog_models.CoverImage

    image_link = Sequence(lambda n: f'https://www.example.com/image/
{n}')
class BlogFactory(DjangoModelFactory):
    class Meta:
        model = blog_models.Blog
    title = Sequence(lambda n: f'Blog {n}')
    content = Sequence(lambda n: f'Blog content {n}')
    author = SubFactory(AuthorFactory)
    cover_image = SubFactory(CoverImageFactory)
```

Now, we will use `BlogFactory` for our API tests:

```
class BlogTestCase4(APITestCase):

    def test_total_blogs(self):
        blogs = BlogFactory.create_batch(4)
        url = '/blog/unpaginated/'

        resp = self.client.get(url, format='json')

        self.assertEqual(len(resp.data['blogs']), 4)
```

In our `BlogTestCase`, we create a batch of four blogs using `BlogFactory`. It is important to understand that `factory_boy` automatically creates the dependent `Author` and `CoverImage` objects for each corresponding `Blog`. Now, developers can focus on writing core business logic test cases, instead of spending too much time creating database objects before they start writing core test cases.

`factory_boy` has a lot to offer, and we have just touched the tip of its iceberg. Explore the official documentation to get a deeper understanding and see different use cases: `https://factoryboy.readthedocs.io/en/stable/introduction.html`.

Using setUpTestData

Using `factory_boy` to create database objects is efficient and recommended. It is possible that we might need to create the same object for multiple tests written for the same API. It's tempting to use the `setUp` method to improve our code quality. The `setUp` method runs before every test case and also deletes any object previously created from a test case. The `setUp` method is executed for every test method in a class, which can cause performance degradation when we have thousands of test cases. Let us look at an example to understand the problem; every time a test runs, we will create the blog object and then delete it before moving to the next object:

```
class BlogTestCase5(APITestCase):
    def setUp(self):
```

```
        self.blog = BlogFactory(title='a1', content='a')
        print('Running Setup multiple times')

    def test_word_count_1(self):
        expected_count = 3
        print('test_word_count_1 running')
        self.assertEqual(expected_count, 3)

    def test_title_length_1(self):
        expected_length = 2
        print('test_title_length_1 running')
        self.assertEqual(expected_length, 2)
```

This will make our test case spend longer in the creation and deletion of blog entries. By adding the print statement, we can verify that the setUp method is called twice because we have two test cases.

We can optimize this by moving all our common data creation code to the setUpTestData class method:

```
class BlogTestCase6(APITestCase):
    @classmethod
    def setUpTestData(cls):
        cls.blog = BlogFactory(title='a1', content='a')
        print('Running Setup only once')

    def test_word_count_2(self):
        expected_count = 3
        print('test_word_count_2 running')
        self.assertEqual(expected_count, 3)

    def test_title_length_2(self):
        expected_length = 2
        print('test_title_length_2 running')
        self.assertEqual(expected_length, 2)
```

In our example, we have moved the blog object creation to the setUpTestData method. By doing this, we create the blog object only once when the test class is instantiated, and the blog object is stored in the test class instance. Our test case would behave in the same manner – that is, after every run of a test case method, the blog object will be reset to the previous state. Under the hood, Django will perform the class-level atomic transaction for every test case run and roll back after execution of the test case, making sure that data returns to its original state after every test case run. For more details, you can refer to the Django official documentation: https://docs.djangoproject.com/en/stable/topics/testing/tools/#django.test.TestCase.setUpTestData.

This optimization might look like a small gain, but as a project matures, such micro-gains help CI/CD pipelines run faster.

Using mocks while writing test cases is very common in the industry, but it is highly recommended to avoid mocking as much as possible. Let us learn about the pitfall of mocking and why we should avoid the use of mocks.

Using mocks as little as possible

While writing test cases, you are bound to come across mocks. **Mocking** is one of the most popular concepts in the testing domain that developers use to write unit/integration tests, and as the application logic/code becomes complex, developers tend to use mocks in complicated pieces of code.

Mock is a built-in package in Python's `unittest` framework. While building web applications, we will have complicated logic or external system integrations, or code blocks that interact with different parts of the application. Writing test cases for such code blocks is difficult and might need a lot of effort. Using `mock`, we can replace some parts of a system we want to test with mock objects. Mocks are beneficial for cases where our code interacts with external systems, or our application code needs a lot of effort to initially set up. We shall not focus on explaining how to use mocks, since it is such a vast topic, and we would need a separate chapter to explain the concepts and use cases of mocks.

Read more in the official documentation for the Python `unittest` about the concepts of mocking in the Python test library: `https://docs.python.org/3/library/unittest.mock.html`

> **Opinionated thought**
>
> Often, developers tend to chase code test coverage metrics, writing tests just for the sake of meeting these. The easiest way to increase coverage for any complex code is by writing unit tests that will mock complex logic; this causes a lot of brittle tests or pseudo tests that forfeits the original use case of writing tests for systems. It is advised to avoid using `mock` while writing test cases as much as possible, writing actual tests that portrays actual production systems. This way, you are not chasing coverage metrics just for the sake of numbers, and the test code coverage percentage represents the real state of the code.

In real life, in production, we see extensive use of `mock`. So, it is recommended that developers make a judgment on when to use it and also understand the deep-rooted problems and concepts of mocking to write better code. Here are a few resources I recommend to anyone who wants to learn more about how to write test cases using mock:

- `https://www.youtube.com/watch?v=ww1UsGZV8fQ`: Explains what goes under the hood when we patch a function while writing tests

- `https://www.youtube.com/watch?v=rk-f3B-eMkI`: Explains how we can stop using mocks that can help us rethink the way we write application code that is well-designed

- `https://www.youtube.com/watch?v=Ldlz4V-UCFw`: Explains different approaches to writing tests without overusing mocks

In the next section, let us learn how to write test cases for our Celery application. Celery is an integral part of our Django application, so it is always recommended to write test cases for the business logic for our celery tasks.

Writing tests for Celery

In *Chapter 8*, we learned how to write code for long-running code blocks in Django using Celery. Celery is an integral part of the business use case, so we should have test cases that cover Celery tasks. A Celery task is nothing but a regular function that has an additional decorator, converting the function to a Celery task.

There is a section in the official documentation for Celery that mentions how we can write tasks in Celery: `https://docs.celeryq.dev/en/stable/userguide/testing.html`. The official documentation also recommends against the use of `Eager` mode for any unit testing. However, at times, using Eager is a better approach than mocking Celery tasks as suggested by the official documentation. Hence, writing tests for Celery tasks is very much a choice, and developers need to make a decision on whether to use mocking or Eager mode, after careful consideration of their project and expertise.

> **Opinionated thought**
>
> Since Celery tasks are regular tasks, most of the time I recommend writing more unit tests that assert the business logic of the Celery task. To test any task initiation, I recommend simply mocking the task call and verifying whether the mocked task is called as per the expected behavior.

Apart from Celery tasks, we also learned how to write decouple our code using the Django Signals framework in *Chapter 7*. In our next section, we shall learn how to write test cases for Django signals and receivers.

Writing tests for signals and receivers

Signals are an important feature provided by Django. Django signals provide us with an abstraction layer, which is a primary reason it is tricky to write test cases for signals and receivers. Since receivers are standalone functions that get triggered on any signal, it is easy to write unit tests for them by writing unit tests for the `receiver` function.

There is no recommended way in the Django documentation that says how you should test signals. Hence, there are multiple ways different developers have found to write tests. One of the approaches I generally use to test receiver connections is disconnecting the receiver functions we want to test, and then verifying whether the receiver got disconnected or not. What does this mean? Let us take an example, `my_handler` receiver connected to the `post_save` Django model signal. Create a new file, `author/receivers.py`, with the following code:

```
from django.db.models.signals import pre_save
from django.dispatch import receiver
from author.models import Author
```

```
@receiver(post_save, sender=MyModel)
def my_handler(sender, **kwargs):
    print('my handle receiver')
```

Now, we want to test that our receiver function is successfully connected to the post_save signal of Author. We shall use the inversion principle to test this scenario; a receiver can only be disconnected if it is connected to the right signal – to use an analogy, you can turn off a bulb only if the bulb is turned on. Add the following code to author/tests/tests_signals.py file:

```
class SignalsTest(TestCase):
    def test_connection(self):
        result = signals.post_save.disconnect(
            receiver=my_handler, sender=Author
        )
        self.assertTrue(result)
```

Read more about the disconnect function in Django's official documentation: https://docs.djangoproject.com/en/stable/topics/signals/#disconnecting-signals.

One more important concept with tests and signals is stopping Django signals for test cases. By disabling signals whenever they are not needed during our test cases, we can run our test cases faster. For example, if we have written application logic where creating a User object automatically triggers model signals that creates more database entries and perform database operations. If we write a unit test that is not going to test or impact the logic change caused by signals, then automatically, our tests will run slow.

To solve this problem, we can use factory_boy to mute all the signals related to Django models or any additional signals. We can use context manager with the mute_signals interface provided by factory_boy to make sure none of the signals are fired when we create a new model object:

```
from factory.django import mute_signals

class SignalsTest(TestCase):

    def test_demo(self):
        with mute_signals(signals.pre_save, signals.post_save):
            # pre_save/post_save won't be called here.
            return SomeFactory(), SomeOtherFactory()
```

Read more here on how we can use factory_boy to mute signals and have faster test execution: https://factoryboy.readthedocs.io/en/stable/orms.html#factory.django.mute_signals.

We have learned how to write test cases and how to test different features of Django. Now, we shall learn about one of the advanced features of testing, Django runners. Django runners are advanced features that can be used by developers to set up a test database or customize test.

Using Django runners

When we execute the Django management command to run tests, `python manage.py test`, Django internally finds the configuration to run tests. This configuration is present in the Django test runner, the `DiscoverRunner` class. Just like any other Django interface, `DiscoverRunner` can also be extended as per requirements.

Let us extend the default `DiscoverRunner` in this section, which should help with any custom use case developers have. Let us imagine that we want to add data to a database before we start the execution of any of the test cases, and then we can use `DiscoverRunner` to load our data. We can create a new class – for example, a `FillData` class, which has a `setup_databases` method that populates the data needed for the test cases. Then, create a `CustomRunner` class that inherits both classes, `FillData` and Django's default runner, `DiscoverRunner`:

```
from django.test.runner import DiscoverRunner

class FillData:
    def setup_databases(self, *args, **kwargs):
        temp = super(FillData, self).setup_databases(*args, **kwargs)
        print("### Populating Test Cases Database ###")
        # Create any data
        print("### Database populated ############")
        return temp

class CustomRunner(FillData, DiscoverRunner):
    pass
```

Once we create the `CustomRunner` class, we need to save it in a given file – for example, `myblog/ backend/common/custom_runner.py`. Now, we need to tell Django to use our `CustomRunner` every time for test case configuration. To do this, we need to set the `TEST_RUNNER` flag in our settings file:

```
TEST_RUNNER = 'common.custom_runner.CustomRunner'
```

Now, every time we run the test management command, Django will pick the test configuration from the `CustomRunner` class we defined. To verify the integration, we can run `python manage.py test`, and we will see the output, which has the `print` statement we added to our `CustomRunner`.

> **Note**
>
> The `CustomRunner` example we shared is an advanced use case, and we expect you to have a good understanding of MRO and other advanced concepts of Python multiple inheritances.

Here are a few custom use cases that developers can configure for Django custom test runners:

- Configuring `parallel` to Django can use to run tests in parallel
- Configuring the test name pattern using `test_name_patterns` to run tests efficiently
- Setting up `tags` and `exclude_tags` to group different tests together
- `failfast` stopping the execution of the test cases immediately when any one of the test cases fails

Django provides a lot of flexibility to create custom runners; however, not all the customizations are generally used by developers. Creating a custom runner is beyond the scope of this book, and we leave it to you to explore the official documentation to get more understanding. Read more on the topic in the official Django documentation: `https://docs.djangoproject.com/en/stable/topics/testing/advanced/#defining-a-test-runner`.

Now that we have covered all the use cases of writing tests in Django, let us move on to the next section, where we will cover best practices to write test cases.

Learning best practices to write tests

So far, we have explored all the advantages of writing test cases and also how we can write them for different Django components. In this section, we shall learn about best practices to write test cases.

Using unit tests more often

Writing test cases is always tricky, and it is very difficult to write test cases for all scenarios. As applications become complex, developers may not be able to write test cases to cover all scenarios, so the best approach to tackle this problem is by writing maximum unit tests. Whenever we are writing new business logic, we generally add small functions/methods that will have new/updated business logic, so writing unit tests for such functions helps to add more coverage. If we have written modular code, then it is easy to write unit tests for our application, and this is something that developers need to understand – *writing test cases is difficult only when the application code is not modular and structured badly.*

Now the question comes, when and how much should we write *integration tests*? The answer is, *it depends.* With more complex business applications, it is better to cover 90% of success scenarios and 30% of failure scenarios by integration tests. For non-critical systems, having fewer integration tests also works, but we should have a couple of success-and-failure-scenario integration tests.

Avoiding time bomb test failures

Time bomb tests are one of the most rookie mistakes developers make. When writing test cases, developers might hardcode the current datetime to the expected datetime or write some logic that would not consider the timezone. This would fail the test cases during midnight or specific times of the day/night.

There is a popular library, *Freezegun* (`https://github.com/spulec/freezegun`), that helps us mock a datetime efficiently to avoid any timebound test case failure. Let us look at an example where we don't want users to spam our blogging application, so you want to limit users to posting 10 blogs per day. We need to write a test case for such a scenario where we can mock any function trying to fetch the datetime. Freezegun can help us mock any datetime automatically within the decorator or context manager scope. In the following code, we mock the datetime called `today` and then change it to the next day. This helps us to write test cases with a cleaner approach:

```python
from django.utils import timezone
from freezegun import freeze_time

from blog.factoryboy import BlogFactory, AuthorFactory
from blog import public

class BlogTests7(APITestCase):
    def test_blog_time_block(self):
        today = timezone.now().date()
        tomo = timezone.now().date() + timezone.timedelta(days=1)

        with freeze_time(today) as frozentime:
            # Post 10 blogs
            author = AuthorFactory()
            create_10_blogs = BlogFactory.create_batch(10,
author=author)
            # Check if the user can post more blogs today
            user_can_post = public.check_if_allowed_to_publish_
blog(author)
            # Validate that the user cannot post a blog today.
            self.assertFalse(user_can_post)

            # Move the date to the next day
            frozentime.move_to(tomo)
            # Validate that the user cannot post a blog tomorrow.
            result = public.check_if_allowed_to_publish_blog(author)
            self.assertTrue(result)
```

Mocking a datetime is one of the most tricky scenarios, but `freezetime` makes it much simpler.

Avoiding brittle tests

As we write tests for our application code, we will come across scenarios where any update to the application code logic breaks the unit/integration tests. This is inevitable, and this is a good sign that you have written good test cases that break when the expected behavior of the code is changed, making the test case fail. However, as a project matures, this becomes a bottleneck, and at times, we might see that updating the application code takes one hour, but fixing test cases may take one full day. This is when you need to take a step back and re-evaluate your test cases and how developers write test cases. There is no silver bullet to fix this issue, and it takes experience to avoid writing brittle tests. However, there are a few guidelines that you can follow to improve:

- **Write more unit tests**: By writing more unit tests that check the core business logic, developers will be able to easily identify failed test cases and the reason behind them. This will help them to fix them faster.

- **Test with dynamic data**: Using dynamic data will help in having more flexible test cases. For example, using `factory_boy` to assert model objects from the return value is always preferred. *Remember that dynamic data for tests is a double-edged sword, and if not implemented appropriately, it can cause a lot of damage.*

- **Avoid business logic coupling**: We should write tests that test only the output of function/methods; if you start testing the implementation details and business logic, then your tests are bound to fail on every update of business logic even though the output contract is consistent.

- **Create small test cases**: It is tempting to add test cases that check multiple use cases for a function/method, but this causes the problem of identifying which test case fails, and also too many updates are needed to identify and solve the failed test case.

These are subjective views and should only be used as advice/guidelines; every developer has a better knowledge of their system and should make a better judgment call on how they can improve writing better tests.

Using a reverse function for URL path in tests

While writing tests in Django/DRF, we will have to use URLs to access our views and API. It might be tempting to use the URL directly as defined, but it is recommended to use the `reverse` function from the `django.urls` package to have robust test cases that manage any update to the URL path. The `reverse` function takes care of the creation of dynamic URLs that can help you focus on writing business logic test cases and not worry about creating URLs. This is also a double-edged sword that can cause failures if the URL path is updated for legacy code, but such changes should be flagged during a code review before your code reaches production.

Using authentication tests

You can easily add authentication for any API in DRF/Django; similarly, it is equally easy for any developer to remove authentication from an API, so it is highly recommended to write authentication validation test cases for each API. We discussed previously how we can write test cases for authentication in the *Testing API authentication* section. Always make sure to write a minimum of one test that verifies the *token-based* or any other authentication system for the API. For example, write one test that will utilize the `self.client.credentials` method to make the login request.

Using test tags to group tests

As our project becomes mature, we will have test cases that are fast, slow, core, or some different type. It might not be possible to run all the tests locally or in the development phase, so we might want to group tests to execute/skip them. We can group different test cases into segments and run them as per the requirement. Django provides a simple decorator interface that can help us tag similar test cases in a group. For example, in the following code, we tag different tests in `SampleTestCase` as `fast`, `slow`, or `core`. This can help us execute all the `core` test cases or `fast` test cases together. Add the following code to the `blog/tests.py` file:

```
from django.test import tag
from rest_framework.test import APITestCase
class BlogTests8(APITestCase):
    @tag("fast")
    def test_fast(self):
        print("fast test running")

    @tag("slow")
    def test_slow(self):
        print("slow test running")

    @tag("slow", "core")
    def test_slow_but_core(self):
        print("slow but core test running")
```

To run a particular group of tagged test cases, we can use the following command:

```
python manage.py test --tag=core
```

This command will only run all the test cases that are tagged with the `core` tag – in our case, `test_slow_but_core`. For more details on how to efficiently use the `tag` concept for test cases, check out the official documentation for Django: `https://docs.djangoproject.com/en/stable/topics/testing/tools/#tagging-tests`.

Using Postman to create an integration test suite

Postman is a great tool to verify the API contract. However, very few developers utilize the power of Postman test scripts to write E2E tests. We can write E2E tests to verify the contract of every API used. This can be achieved by writing test scripts. Once we have the test scripts, we can write a monitor to continuously test the results. This way, we can set up an E2E test setup in any of our preproduction environments, and we will be notified of any of the test failures. Here are a few important links that can help you get started with writing the Postman E2E test:

- How to write test scripts in Postman: `https://learning.postman.com/docs/writing-scripts/script-references/test-examples/`

- How to create monitors that will periodically check for the tests to pass: `https://learning.postman.com/docs/monitoring-your-api/intro-monitors/`

- How to use the Postman CLI to run test cases from the CI/CD pipeline: `https://learning.postman.com/docs/postman-cli/postman-cli-overview/`

Writing Postman test scripts and periodically running them via the Postman monitor is one of the simplest and fastest ways to set up any test suite for your APIs. You can also use any other tool that is available on the market for such a purpose – for example, Insomnia provides such a testing interface (`https://docs.insomnia.rest/insomnia/unit-testing`).

Creating different types of tests

Writing tests for application code is a part of a development project, and teams should always calculate the time taken to write test cases and then provide a final estimate of the project. Apart from writing unit tests and integration tests, we should write a few system-level tests that help to maintain the good health of the code quality and system:

- **Coverage threshold**: We should have a code coverage threshold for each **Pull Request** (**PR**). Whenever the raised PR decreases the overall coverage, we should mark our CI pipeline as a failed test.

- **Package file tests**: Third-party packages are used extensively in production, and we should be well aware of what code is going to production. When a developer adds a particular package to the repository, they will extensively test the package and its API before moving the code to production. The testing of the package is always based on the then-available version of the package, but if the package gets updated, then it is highly likely that the code can behave in some unexpected way. To avoid this, it is always recommended to write tests that check whether the Python package added in the `requirements.txt` file has the specific version.

- **Linting tests**: Automatic linting is one of the most important setups that you should add to your projects. Linting should be configured using **githooks**, like `commit` and `push`. It is also possible that the local *githook* is broken or the developer bypassed the linting test in their local. To avoid such scenarios, we should have linting tests set up in our CI pipeline that will fail whenever a linting test fails. For linting, we can add `flake8` (`https://pypi.org/project/flake8/`) and `black` (`https://pypi.org/project/black/`). In *Chapter 12*, we will discuss in detail their integration.

- **Migration file tests**: Migration is one of the most-loved features of Django, which takes care of all database-related changes. It is easy to perform any database-related change directly from the application code. In *Chapter 2*, we learned to always write backward-compatible migration. To enhance this practice, we should add tests that check for any migration change that performs a destroy operation. For example, we should have tests that fail whenever there is a new migration file that contains a "drop table" command or any other destruction command. This way, there is no accidental destructive operation that goes to production without manual intervention.

- **Verify test filenames**: Django/Python automatically detects all the test cases from the filename pattern – that is, files starting with `test`. However, developers may make a typo in the filename or add the `test` keyword to the end of the filename. In such cases, tests will not be detected by the framework, and tests will not be executed automatically. Hence, we should write a test that will detect all such filenames that have mismatches.

These system-level tests are added using **githooks** and **CI pipelines**, and we shall discuss more about them in *Chapter 12*.

Avoiding tests

When should we avoid tests? Although it is always better to add tests to every project, we all know it is not possible to always write test cases. Even when we are having a real-time crunch, I would still recommend writing a few basic integration test cases that give some validation and confidence in a project.

Here are a few scenarios when you may decide to skip writing test cases, although these are subjective:

- **A hobby project**: If you are doing a hobby project and you are sure the project is not going to be scaled. However, be mindful that a lot of hobby projects have become multi-billion-dollar companies.

- **A solo developer with a simple use case**: If you are a solo developer of a simple project that only performs CRUD operations or something very small, then you can avoid writing tests.

- **Proof-of-concept projects**: Proof-of-concept projects are short-lived and are in the exploration phase; they can avoid test cases because the project can pivot easily.

Most developers write test cases after the application code is done, and during time- critical in start-up, we might end up pushing code to production without writing test cases. Avoiding test cases can be tempting but should not be encouraged at all.

There is a development process that emphasizes writing test cases first and then writing an application. Let us learn more about **test-driven development** (**TDD**) in the next section.

Exploring Test-Driven Development

TDD is a type of development process where we first write the tests that will fail, followed by writing our application code, which will make all the failing tests pass. This helps us to write smaller code blocks and also have a set contract on what each function/method should provide as output. TDD is a thought process, and new developers might need some handholding to shift to this mind space.

We shall not go into the details of TDD; there is a course that explores this in detail: `https://www.packtpub.com/product/hands-on-test-driven-development-with-python-video/9781789138313`.

With this, we wrap up all the topics related to writing tests in Django/DRF.

Summary

In this chapter, we learned different testing techniques and the value that tests add to the development cycle and production stability. We explored how to set up tests in DRF and write test cases for different scenarios. We learned how to use third-party packages such as `factory_boy` to improve the developer experience of writing tests.

We saw a few best practices and recommendations to improve test cases in the development cycle. Postman as a tool can be very helpful in writing tests to monitor contracts; we learned how we can use Postman features to write integration tests for our APIs.

In *Chapter 10*, we will discuss the best practices/conventions that you can follow while developing applications in Django and DRF. Conventions and good practices are mostly subjective and come from years of experience in writing production-level code. I am also going to share my views and the lessons I have learned from working on different projects related to Django in the next chapter.

I would like to especially thank *Adam Johnson* for writing the amazing book *Speed Up Your Django Tests*, which inspired me to cover a few additional topics about testing in this chapter.

10

Exploring Conventions in Django

In previous chapters, we learned about the different features of Django and how you can utilize these features to build a scalable Django application. Django is considered to be opinionated; it doesn't give a lot of flexibility to developers. Being opinionated has led to a love-hate relationship between Django and the developer community, where a lot of developers love the fact that Django has a particular way of performing certain implementations and developers do not have to think too much about how to use Django to build a web service. However, many developers don't like Django due to the lack of flexibility and its being too rigid about how a particular code should be written.

In this chapter, we will learn about certain conventions you can follow while working with Django. You should note that this chapter contains a lot of opinionated concepts that I have learned while working on numerous Django projects and consulting for different companies.

Let us look into the main topics we will be covering in this chapter:

- Code structuring for Django projects
- Working with exceptions and errors
- Using feature flags
- Configuring Django to deploy to production

Technical requirements

We will learn the conventions of different Django topics we have covered so far in this book. You should be well versed in all the topics explained in earlier chapters.

Here is the GitHub repository that has all the code and instructions for this chapter: `https://github.com/PacktPublishing/Django-in-Production/tree/main/Chapter10`.

Code structuring for Django projects

When we start a new project in Django, the Django management command creates the basic files needed to create our Django project. In *Chapter 1*, we learned how we can set up our Django project and create different Django apps. In this section, we shall collate all the information we have gathered in previous chapters on using different files for different purposes in Django.

Creating files as per functionalities

As a first step in bifurcation, we have already learned that we should create multiple Django apps in our Django project. The splitting of a Django app depends on the business logic functionalities; for example, our blogging project can have `blog` and `author` as two different Django apps. Inside each app, we would split our code into multiple files. Here is the recommended list of files we should consider while working on Django:

- `urls.py`: Each Django application should have a `urls.py` file that contains all the routing logic for the given application.
- `api.py`: We should add a definition for all the APIs in the `api.py` file and link it to `urls.py`. A lot of developers also use `views.py` for the same purpose.
- `constants.py`: We should use the `constants.py` file to add all the constants used in our Django application.
- `models.py`: We should put all our database table definitions in our `models.py` file. Django already mandates the use of a `models.py` file for any database definition-related code.
- `serializers.py`: Add all the DRF serializer logic inside the `serializers.py` file. Avoid adding serializer logic inside the model file to maintain a separation of concern.
- `common.py`: A lot of the time, we might see the repetition of logic in the same Django app. It is better to create a `common.py` file that contains all the common logic used by the application.

- `utils.py`: As we build applications, we write a lot of utility functions to perform certain operations. For example, if we wanted to trim whitespace from an article, we would create a `util` function for this purpose and put that function in the `utils.py` file so that it can be reused. The main difference between `common.py` and `utils.py` is that `common.py` can be dependent on other files, such as `models.py`, but `utils.py` should not be dependent on other files.

- `public.py`: As our project matures, we shall observe interdependency between different Django applications. It is important to maintain a hierarchy in files to have a smooth developer experience. We should keep all functions and classes that need to be accessed by other Django apps in the `public.py` file.

- `_private.py`: We should keep all the code that should be accessible only by the internal Django app in the `_private.py` file. The primary difference between `public.py` and `_private.py` files is that public file functions can be accessed by any other Django app and private functions should only be accessed by files in the same Django app. Public functions can internally call private functions and they can have a wrapper around the private function to expose the function to another Django app.

- `tasks.py`: We learned how to use Celery in *Chapter 8*. As mentioned in that chapter, we should put all our Celery task code in the `tasks.py` file.

- `signals.py`: Add all the custom Django signal definitions to the `signals.py` file.

- `receivers.py`: Add all the Django receiver-related code to the `receivers.py` file.

- `admin.py`: Out of the box, Django provides a very useful admin interface. All the code related to Django admin should be inside the `admin.py` file.

- `tests`: In *Chapter 9*, we saw how we should utilize `factoryboy` and `tests` files to write tests in Django. All the tests should go inside the `tests` folder.

We have discussed different types of files that are used inside a Django application. It is not mandatory to create all these files right from the start, but consider it as a reference to build a large-scale maintainable Django application.

As the project matures, you will also see that maintaining code becomes difficult, even when having separate files as mentioned in the preceding list. When such a situation occurs, you should decide whether you should break your Django application into smaller sub-applications or break individual files into modules. For example, if your `public.py` file has 500+ lines, then you should consider converting the file into a module, by creating a `public` folder and then adding `__init__.py` to it. Once you convert the file into a module, you can start breaking the `public.py` file logic into multiple smaller files. This can be applied to any of the files mentioned in this section.

> **Note**
>
> From my experience in the industry, I have learned to follow the rule of 500 lines: if any file contains more than 500 lines of code, it is recommended to convert the file into a module and split the logic into multiple smaller files.

In the next section, we will learn how to avoid circular dependencies and how different files can help us plan better to structure our code.

Avoiding circular dependencies

Circular dependencies are bound to occur as projects grow. There is no single guaranteed approach via which circular dependencies can be avoided. Python automatically gives us a hint that part of our code has a circular import error. Here is an example of a circular import error thrown by Python:

```
ImportError: cannot import name '<name>' from partially initialized
module '<module>' (most likely due to a circular import)
```

But we can follow certain guidelines to save us from this common issue:

- Never import models from a different app. If needed, use the `public.py` file interface to import models of different applications. This also helps in abstracting out the implementation details of different apps. For example, if at a later point in time there is any modification made to the table, then we do not have to update the code of other Django apps; rather, we have to guarantee the contract of the `public.py` interface.

- Always try to dispatch signals and use receivers to communicate between two Django apps. This helps to avoid the direct import of other app code.

- If we still get circular imports, the easiest solution is to break the culprit file into two different sections and make sure they are not calling one another directly.

Circular dependency is a tough problem to solve, and most of the time, a custom approach would be needed to solve it. The concepts we have learned in this section are only an outline that can support you in planning your code according to a style guide and structure it better.

Creating a "common" app

We will create a Django app named common. This app will contain all the common code that can be used across the project. For example, if we have a common model that is used for saving simple configurations, then we should add the given app to the common app. The common app can be imported by any other Django app, but it should never import any other Django app.

For example, if we need to create a simple *key-value store table* that can be used by multiple Django apps to fetch configurations, we can create the key-value store in the common app.

Here is a list of features that I have used in the common app across different Django projects:

- Holding all the generic custom exception classes so that other Django apps can use them.

- Including custom pagination classes that are common across the project.

- Adding common logging functions so that all logs are passed through just one function. This helps in maintaining logs at a large scale and also having the flexibility to update logging implementation without breaking the existing system.

- Creating custom middleware.

- Holding base model classes. For example, each model should have a `created_at` and `updated_at` field. Create a base model and put it in the common app so that all Django app models are inherited from this base model class.

- Using base admin and custom admin mixins.

Though the common app is super helpful and helps us follow the **Don't Repeat Yourself** (**DRY**) principle, we should also be mindful of what code we are putting in the Common app module. As projects expand, it might be a good idea to revisit the common app and decide whether it makes sense to create a separate app and move all the code logic to it.

Now, let us learn how we can set up different environments using the Django settings.

Working with a settings file for production

When we create a new Django project, the Django management command creates a `settings.py` file that contains all our Django configurations. But as our project grows, we will have different configurations for different environments, such as *local*, *dev*, *stage*, and *prod*. We should start maintaining different files for different environments.

A recommended approach to solve this problem is to create the following file structure:

- `settings/base.py`: This file would contain all the common Django settings that would be applicable for all the environments, for example, `INSTALLED_APPS`.

- `settings/local.py`: This would contain all the configurations to run Django for local development. We need to import all the settings in the `base.py` file to `local.py`.

- `settings/prod.py`: This would contain all the configurations to run Django in production. Similarly, we would import all the settings from the `base.py` file.

We can create multiple settings files depending on our requirements. Now that we have split our settings file into multiple files, we need to tell Django to pick the appropriate settings. There are multiple ways to configure the Django server to pick a settings file:

- Using `DJANGO_SETTINGS_MODULE`. We can set the settings file path to the environment variable from the command line. For example, for Unix, we would use this command:

  ```
  export DJANGO_SETTINGS_MODULE=config.settings.local
  ```

- Pass the settings file using the `--settings` flag while starting the server, using the following Django management command:

  ```
  python manage.py runserver --settings=config.settings.local
  ```

We should also remember that we do not hardcode or commit any of the secrets to our code repository. Rather, we should always use environment variables to import the secret value on the fly. This is particularly important when we deploy code to production. Python has an `os` library that can be used to import the environment variables. For example, `os.environ.get("SQL_USER", "user")` would load the database user from the `os` environment variable:

```
DATABASES = {
    "default": {
        "ENGINE": os.environ.get("SQL_ENGINE"),
        "NAME": os.environ.get("SQL_DATABASE"),
        "USER": os.environ.get("SQL_USER", "user"),
        "PASSWORD": os.environ.get("SQL_PASSWORD", "password"),
        "HOST": os.environ.get("SQL_HOST", "localhost"),
        "PORT": os.environ.get("SQL_PORT", "5432"),
    }
}
```

We have learned about the importance of structuring code in the Django project and how it can benefit us. In the following section, we shall see how we should manage exceptions and error handling in the Django project.

Working with exceptions and errors

Developers should always think through all the corner cases possible and write code to handle all those corner cases. Unfortunately, bugs, exceptions, and errors can still slip into production, leading to our users having a bad experience. We have already learned, in *Chapter 9*, how test cases can help us catch these errors better in the development phase itself, and to improve the user experience, we should show appropriate error messages whenever something breaks. For example, if the server is expecting `name` to be present in the body of the request and it is missing, then we should send an error message explicitly saying `name field is missing`.

Here are a few important points we should follow while handling exceptions and errors:

- Always send the appropriate status code. If the request body is missing certain information, then pass 400 and then pass a message explaining the missing information.

- If the error is an authentication error, then send 401. If we use DRF, then DRF automatically responds with the appropriate HTTP status code.

- If the request error is an authorization error, then send 403.

- Create a structure for the error code. For example, if an API in the blog app is getting bad requests, we should create a custom code, BLG0023 or BLG0042, that helps to identify which part of the code is breaking and returning a bad response. In our example, with BLG0023, we can identify more details about the service, module, and functionality using the API error code. This is one of the most widely used strategies in the industry; for example, you might have noticed such error codes displayed in ATMs.

- Integrate error monitoring tools such as Rollbar (https://rollbar.com/) or Sentry (https://sentry.io/welcome/) to capture errors. We shall discuss the integration in detail in *Chapter 14*.

- While implementing exception blocks in code, we should avoid blanket exception handling and silently ignoring errors. There is a detailed Stack Overflow answer that explains the reason behind this: https://stackoverflow.com/questions/21553327/.

Let us take an example of how we should implement better error handling in our code base. Suppose a user requests a particular blog that is not present in the database. We should handle such a scenario as follows:

```
@api_view(['GET'])
def get_blog_by_id(blog_id):
    try:
        blog = Blog.objects.get(id=blog_id)
    except Blog.DoesNotExist:
        return Response(
            status=status.HTTP_404_NOT_FOUND,
            data={'error': 'Blog does not exist',
                  'error_code': 'BLG0012'}
        )
    blog_data = BlogSerializer(blog).data
    return Response({'blog': blog_data}, status=status.HTTP_200_OK)
```

Here, we are capturing the error and sending an appropriate error message and error code for the team to debug, along with the correct HTTP status code.

We learned about error handling and monitoring in this section. As we build new features in our project, we want to gain more confidence in our feature but also have options to revert any breaking changes as soon as we detect them. This is where feature flagging/the kill switch come into the picture.

Using feature flags

A feature flag is a technique to enable or disable certain functionalities in code during runtime without deploying new code. This helps developers to toggle different features in production at a much faster rate. For example, if a company has incorporated Stripe and PayPal payment gateways for payments in their app, but for some reason Stripe is down, then the team can quickly change their payment gateway to PayPal without any code change or deployment. This is very common for critical systems and also while building new systems.

In the following example code, we are storing the payment gateway configuration in our `KeyValueStore` database table. This helps us to control the payment gateway at runtime. `KeyValueStore` is a database model that is stored in the `common/models.py` file:

```python
class KeyValueStore(models.Model):
    key = models.CharField(max_length=255, unique=True)
    value = models.JSONField()
    def __str__(self):
        return self.key
```

Now, create a public interface that can be used by other apps to get the config value. Create a `common/public.py` file with the following values:

```python
from common import models

def get_current_config(key_name):
  try:
    return models.KeyValueStore.objects.get(key=key_name).value
  except models.KeyValueStore.DoesNotExist:
    return {}
```

Now, we can use the `common` config function to access any of the values stored in our `KeyValueStore` model:

```python
from common import public
def get_current_gateway():
    return public.get_current_config('payment_gateway')

def get_payment_gateway():
    payment_config = get_current_gateway()
    if payment_config.get('name') == 'stripe':
```

```
        return 'stripe'
    elif payment_config.get('name') == 'paypal':
        return 'PayPal'
    return 'stripe'

def make_payment(request):
    # do something related to payment
    payment_gateway = get_payment_gateway()
    # process payment
```

Feature flagging is very important for developers to learn as it gives them added confidence in the system. Similarly, this strategy helps developers use **kill switches** and also create an A/B testing and rollout strategy without relying on the redeployment of code. It is easy to implement basic feature flagging and kill switches using simple database and code logic, but as systems grow in size, you can think of using external systems for such purposes. Tools such as Flagsmith (`https://flagsmith.com`) and LaunchDarkly (`https://launchdarkly.com`) can help with building complicated feature flags and rollout strategies.

We have learned about feature flagging and how it is important in the development cycle. Now we shall look at a few tips on how to configure Django for production.

Configuring Django for production

In this section, we will learn about a few key configurations we need to set before deploying to production. We will first discuss the configuration we need to update the `settings.py` file:

- Update the `SECRET_KEY` value and pass it via the environment variable
- Set `DEBUG = False` for production so that sensitive information via an error stacktrace is not shown, whenever there is an error in the Django app
- Set `ALLOWED_HOSTS = ["dip.com", "xyz.com"]`, and list all the domains that the Django project would be served on
- Set `APPEND_SLASH = True` so that if any request is missing a slash at the end, Django can add a slash automatically
- Set `TIME_ZONE` appropriately

There are a few more configurations that you can update in the Django settings. They depend on the Django project use case. Apart from the preceding mentioned settings, let's learn about a couple of other configurations we need to perform:

- Change the Django admin path from `/admin` to something that is non-guessable
- Always use `timezone.now()` for any time-related operation

In this book, we have learned how to create REST APIs – unless we are building purely for mobile applications, we would have to configure **Cross-Origin Resource Sharing** (**CORS**), which would enable our Django application to be accessed from other origins.

Setting up CORS

CORS is a mechanism that allows resources to be accessed from different origins. CORS is particularly applicable on browsers rather than other devices, such as mobile or IoT. There is a third-party package, `django-cors-headers`, that is widely used to configure CORS for Django applications. Please remember that CORS and **Cross-Site Request Forgery** (**CSRF**) are different, and *Django provides support for only CSRF but not CORS.*

Here are the steps to configure CORS in a Django project:

1. Install the `django-cors-headers` package by running this command:

    ```
    pip install django-cors-headers
    ```

2. Add `corsheaders` to the `INSTALLED_APPS` config list:

    ```
    INSTALLED_APPS = [
        ...,
        "corsheaders",
        ...,
    ]
    ```

3. Add `corsheaders.middleware.CorsMiddleware` to the `MIDDLEWARE` configuration in the Django settings:

    ```
    MIDDLEWARE = [
        ...,
        "corsheaders.middleware.CorsMiddleware",
        "django.middleware.common.CommonMiddleware",
        ...,
    ]
    ```

4. Now, add the domains/subdomains you want to enable CORS for in the `CORS_ALLOWED_ORIGINS` config in the `settings.py` file:

    ```
    CORS_ALLOWED_ORIGINS = [
        "https://example.com",
        "https://sub.example.com",
        "http://127.0.0.1:9000",
    ]
    ```

We have configured our basic CORS configuration for our project.

> **Note**
>
> There are a few other configurations that can be made, but they depend on the use case of the project. Feel free to check the official documentation to learn about more configurations that can be applied, at `https://github.com/adamchainz/django-cors-headers`.

One last concept we need to learn to deploy our Django application to the project is **Web Server Gateway Interface (WSGI)**.

Exploring WSGI

WSGI is a specification laid out in PEP 333. This specification defines a standard interface between web servers such as Nginx and Apache and Python web frameworks/applications such as Django and Flask. The details of the implementation of WSGI are out of the scope of this book. We will just learn how to integrate WSGI into a Django project.

The boilerplate code Django creates already has a `wsgi.py` file generated. This is the entry point for a Django project in production. We need to add configuration in the `wsgi.py` file to set the `settings.py` file, by setting the `settings.py` file path in the `DJANGO_SETTINGS_MODULE` variable. For example, if our settings file is present in `config/settings/prod.py`, then we should set the path as `os.environ["DJANGO_SETTINGS_MODULE"]="backend.settings.prod"`.

Once we have configured the `wsgi.py` file, we will use `gunicorn` to start the project. Please remember, WSGI is the specification and `gunicorn` is the Python WSGI HTTP server for UNIX. We will not deep dive into the concepts of `gunicorn`, since it is a framework in itself. Rather, we shall focus on the integration of `gunicorn` with our Django project.

> **Important note**
>
> Gunicorn doesn't support the Windows environment. Since we would be using a Linux environment to deploy our code, please skip this step if you are using a Windows setup. Alternatively, you can use the waitress Python package to test locally, but this would not be used in any other section. For more details, please check the GitHub readme of *Chapter 10*.

Here are the steps to integrate `gunicorn` into any Django project:

1. We have already seen how to configure the `wsgi.py` file. You can setup `wsgi.py` as per the instructions mentioned earlier in this section.

2. Now, we shall install `gunicorn` with the following command in our project:

    ```
    pip install gunicorn
    ```

3. Start the `gunicorn` server with the WSGI file. Make sure you are running the following command from the same path where the `manage.py` file lives:

```
gunicorn config.wsgi:application --bind 0.0.0.0:8000 --workers=4
--max-requests=512 --max-requests-jitter=64
```

There are a lot of flags in the code snippet. Let us now break down each config:

I.   `config.wsgi.application` is the path where we have the Django WSGI configuration. This is the entry point to our Django application in production.

II. `--bind 0.0.0.0:8000` is the configuration to bind the application to the `0.0.0.0:8000` port, and we shall configure an Nginx reverse proxy to send the traffic to port `8000`.

III. `--workers=4` is to run four workers of the `gunicorn` server.

IV. `--max-request=512` is to tell `gunicorn` to restart the server whenever it has processed 512 requests. This is particularly helpful when we have memory leaks in the application.

V.   `--max-request-jitter=64` is to introduce a jittering effect while restarting the servers, else we might see all the servers restart at the same time in the background, causing downtime.

Django also supports deployment using **Asynchronous Server Gateway Interface** (**ASGI**), which provides specifications for a standard interface between async-capable Python web servers, frameworks, and applications. The implementation of ASGI is beyond the scope of the book. However, if you want to implement it, you can check the official documentation for further details at `https://docs.djangoproject.com/en/stable/howto/deployment/asgi/`.

We have covered code concepts relating to Django and other packages that are needed to build a scalable Django project.

Summary

In this chapter, we learned about different conventions we can implement in our Django project. We learned how we can create different files in our Django project and structure them properly to write scalable and maintainable code, as well as avoiding circular dependency errors as our project expands. Exception and error handling is an important concept that we have learned about in this chapter. We understood how error codes can help us debug properly in production.

We saw how feature flags can help us to have better control over our code and enable/disable certain functionalities during runtime without requiring any additional deployment of code to our servers. Then, we saw some tips and tricks to configure our Django application before deploying it to production. CORS and WSGI are critical specifications for web development, and we learned about them, as well as the `django-cors-headers` and `gunicorn` packages, in this chapter. We also learned about application development, as well as how to write scalable applications.

In the next part, we will start learning about different tools used for development, like Docker, Git, and more. We shall learn the best practices for using version control, specifically Git and githooks, and how we can utilize **Continuous Integration** (**CI**) pipelines such as GitHub Actions and GitLab CI for a better development process.

Part 3 –
Dockerizing and Setting Up
a CI Pipeline for
Django Application

In this part, we will learn how to integrate Docker into our Django project. We can use Docker to create a platform-agnostic Django application, with easier setup and deployment in production. We will learn how to work with version control software (`git`) and explore how to set up our CI pipeline. We will learn different automations using a CI pipeline and see how we can implement a code review process for our Django application.

This part has the following chapters:

- *Chapter 11, Dockerizing Django Applications*
- *Chapter 12, Working with Git and CI Pipelines Using Django*

11

Dockerizing Django Applications

In previous chapters, we discussed how to write application code for a Django application. In the next few chapters, we will explore the different tools that we can use to improve our development experience, deploy our code to a production system that is scalable, and monitor our production system.

Let us get started with improving our development experience. In *Chapter 1*, while setting up our local development for the Django project, we mentioned **Docker**. We will learn how to use Docker to develop our Django application. In this chapter, we shall primarily focus on Dockerizing our Django application locally with all the services running on our local system using containers.

In the previous chapters of this book, we used different remote services for Postgres and Redis while working with our Django project locally so that we did not get blocked by any installation steps. In order to ensure that everyone can easily follow all the instructions given in the book without encountering difficulties due to potentially outdated tutorials or variations in local system configurations, I made sure to use the bare minimum of local system dependencies so that the guide is easy to follow. But as our project grows, it will become important to do the local setup of our project; for this purpose, Docker is the tool of choice for most developers.

In this section, we shall learn how to set up Django, Celery, Celery Beat, Redis, and Postgres in our local system using Docker. Since the Docker setup of projects is done only during the initial phase of the creation of the project, we shall take a boilerplate project template and Dockerize the project without getting into the details.

Let us look at the main topics we will be covering in this chapter:

- Learning the basics of Docker
- Creating a Dockerfile for a Django project
- Composing services using `docker-compose.yaml`

Technical requirements

In this chapter, we shall discuss how to use Docker with a Django application. We shall not be getting into the details of Docker and containers. We expect you to have a foundational understanding of Docker. The following is expected from you for this chapter:

- An understanding of containers and Docker
- An understanding of how to install Docker on your local system
- Familiarity with accessing Docker development environments

Since we shall not be focusing on the basics and details of Docker, if you are not familiar with the requirements mentioned, please read/revise the concepts of Docker from these books/videos:

- *Docker Deep Dive, Second Edition* – By *Nigel Poulton*
- *Docker for Developers* – By *Richard Bullington-McGuire, Andrew K. Dennis,* and *Michael Schwartz* from *Packt Publishing*
- *The Ultimate Docker Container Book, Third Edition* – By *Dr. Gabriel N. Schenker* from *Packt Publishing*
- *A Developer's Essential Guide to Docker Compose* – By *Emmanouil Gkatziouras* from *Packt Publishing*

Here is the GitHub repository that has all the code and instructions for this chapter: `https://github.com/PacktPublishing/Django-in-Production/tree/main/Chapter11`.

Learning the basics of Docker

Docker is an open source containerization platform that allows developers to build, deploy, and manage containerized applications. Before we get into the concepts of Docker, let us learn a few concepts that will build the foundation:

- **Virtualization** – The concept of virtualization has been around since the 1970s. It is the process of running virtual instances of different operating systems/software in a virtual layer that is abstracted from the actual hardware.

Consider an analogy of 10 bungalows in an area, where each bungalow has its own amenities that cannot be shared with the neighbors. All the owners live in their bungalows and are limited by the area allocated to them. They can do anything in their bungalow but do not have access to their neighbor's bungalow. *Figure 11.1* shows an example of what virtualization means using a virtual machine.

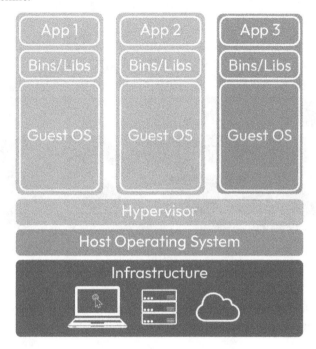

Figure 11.1: Concept of Virtual machine using virtualization

- **Virtual machine** – Virtual machines run on top of a physical machine, have access to the resources allocated to them, and cannot access more resources than their allocation.

- **Hypervisor** – A hypervisor is a program for creating and running virtual machines.

The concept of a virtual machine works seamlessly and is still used in the industry, but it also has a problem with inefficient use of resources. Taking the same analogy, why does every bungalow have its own parking space? They could have a single car park that could be used by residents to park.

Docker solves this problem by letting developers pack their applications into small containers and then the containers can run on the machine independently. Taking the same analogy, Docker can be considered as an apartment building, where tenants stay in each apartment and each tenant can use the amenities. Each apartment has a parking space allocated to them in the parking area. Now we have an efficient way of utilizing the parking area.

Docker provides logical separation between containers, and each container runs independently in its own space without knowing about/accessing the data in other containers (unless specifically shared).

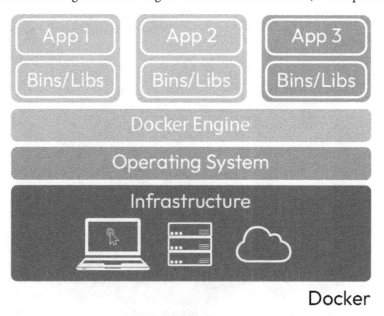

Figure 11.2: Docker high-level architecture

While working with Docker, we come across multiple terms related to Docker. Here are the key concepts of Docker that will help us to understand Docker better:

- **Dockerfile** – A Dockerfile is a textual document providing guidance for constructing a Docker image and detailing the foundational operating system for the container. It outlines necessary elements such as languages, environmental variables, file paths, network ports, and other requirements, as well as the intended actions for when the container is executed.

- **Docker images** – Docker images encapsulate both the executable application source code and all the tools, libraries, and dependencies essential for the application's containerized execution. The act of running a Docker image transforms it into either a singular instance or multiple instances of the container.

- **Docker containers** – A Docker container image serves as a compact, independent, executable software package containing all the essentials for application execution, encompassing code, runtime, system tools, libraries, and configurations.

- **Docker Engine** – Docker Engine functions as a client-server application, facilitating the processes and workflows necessary for constructing, transporting, and executing applications based on containers. It initiates a server-side daemon process responsible for managing images, containers, networks, and storage volumes.

- **Docker daemon** – The Docker daemon is the behind-the-scenes service on the host system responsible for overseeing the creation, execution, and distribution of Docker containers. This process operates on the host's operating system, serving as the interface through which clients communicate.

- **Docker registry** – The Docker registry acts as the storage hub for Docker images. It serves as a repository, residing either locally for individual users or publicly, such as on Docker Hub, enabling collaborative efforts in application development. Teams, even within the same organization, can seamlessly exchange or distribute containers by uploading them to Docker Hub—a cloud repository akin to GitHub.

- **Docker Hub** – Docker Hub, the self-proclaimed "world's largest library and community for container images," functions as the public repository for Docker images. Boasting a collection exceeding 100,000 container images, it amalgamates contributions from commercial software vendors, open source initiatives, and individual developers. The repository encompasses images crafted by Docker, Inc., certified images from the Docker Trusted Registry, and a multitude of others. Docker Hub users enjoy the freedom to effortlessly share their images and can readily access preconfigured base images to kickstart any containerization endeavor.

- **Docker Compose** – Docker Compose simplifies the orchestration of multiple containers, treating them collectively as a single service. While executing each container independently, it facilitates seamless interaction and communication among containers.

We will learn more about Docker Compose in the latter part of this chapter (in the *Composing services using docker-compose.yml* section). Now let us learn how to install Docker and use it on our local system.

Installing Docker

Docker is an open source piece of software available to download for free from the official website. Docker is available for the Linux, Windows, and Mac operating systems. The instructions for installing and running Docker for each platform are well documented and explained in the official guide, hence we shall not give them here. Please follow the instructions given:

- For macOS users, please follow the steps given at `https://docs.docker.com/desktop/install/mac-install/`.

- For Windows users, please follow the steps given at `https://docs.docker.com/desktop/install/windows-install/`.

- For Linux users, please follow the steps given at `https://docs.docker.com/desktop/install/linux-install/`.

Once we have Docker installed, we can run the `docker version` command to verify our installation. *Figure 11.3* shows the output of all the Docker drivers needed to run Docker properly. *Please note: the version numbers can be different for different platforms and depending on when you are installing Docker.*

```
docker version
Client:
 Cloud integration: v1.0.29
 Version:           20.10.22
 API version:       1.41
 Go version:        go1.18.9
 Git commit:        3a2c30b
 Built:             Thu Dec 15 22:28:41 2022
 OS/Arch:           darwin/arm64
 Context:           default
 Experimental:      true

Server: Docker Desktop 4.16.1 (95567)
 Engine:
  Version:          20.10.22
  API version:      1.41 (minimum version 1.12)
  Go version:       go1.18.9
  Git commit:       42c8b31
  Built:            Thu Dec 15 22:25:43 2022
  OS/Arch:          linux/arm64
  Experimental:     false
 containerd:
  Version:          1.6.14
  GitCommit:        9ba4b250366a5ddde94bb7c9d1def331423aa323
 runc:
  Version:          1.1.4
  GitCommit:        v1.1.4-0-g5fd4c4d
 docker-init:
  Version:          0.19.0
  GitCommit:        de40ad0
```

Figure 11.3: Output of the docker version command

Once Docker is installed, let us test Docker on our local system.

Testing Docker on your local system

Now that Docker is installed in our local system, let us try to run a Docker container to verify the Docker setup. Docker has a sample image that we can use to run a container. Docker provides an official image that we can use to test. The `hello-world` image is a simple example that is used by developers to verify the setup.

By running the following commands, we can verify the Docker setup on our local machine:

```
> docker pull hello-world  && docker run hello-world
```

Figure 11.4 shows the output we should be able to see when we run the `docker run` command:

```
docker pull hello-world  && docker run hello-world
Using default tag: latest
latest: Pulling from library/hello-world
478afc919002: Already exists
Digest: sha256:ac69084025c660510933cca701f615283cdbb3aa0963188770b54c31c8962493
Status: Downloaded newer image for hello-world:latest
docker.io/library/hello-world:latest

Hello from Docker!
This message shows that your installation appears to be working correctly.

To generate this message, Docker took the following steps:
 1. The Docker client contacted the Docker daemon.
 2. The Docker daemon pulled the "hello-world" image from the Docker Hub.
    (arm64v8)
 3. The Docker daemon created a new container from that image which runs the
    executable that produces the output you are currently reading.
 4. The Docker daemon streamed that output to the Docker client, which sent it
    to your terminal.

To try something more ambitious, you can run an Ubuntu container with:
 $ docker run -it ubuntu bash

Share images, automate workflows, and more with a free Docker ID:
 https://hub.docker.com/

For more examples and ideas, visit:
 https://docs.docker.com/get-started/
```

Figure 11.4: The output of the hello-world Docker image

Now that we have verified our Docker setup and have run the `hello-world` Docker image, let us learn a few important commands that we need to use.

Important commands for Docker

The Docker CLI has multiple commands that we can use to interact with and perform different operations. Here we shall look into the most common commands that are used regularly by developers:

- `docker ps` shows you the list of running containers along with essential information, such as container ID, names, and status. It's a quick way to check what containers are currently active.

- `docker pull` downloads a Docker image from a registry.

- `docker build` builds a Docker image from a Dockerfile.

- `docker run` creates and starts a new container from a specified image. It allows you to customize various aspects of the container, such as port mappings, environment variables, and more.

- `docker start` starts a stopped container. It brings a container back to life, allowing you to resume its operations.

- `docker exec` executes a command inside a running container. It's handy for performing tasks or running applications within a specific container without starting a new one.

- `docker stop` stops a running container.

- `docker restart` restarts a container.

- `docker log` displays the logs of a container.

> **Read more**
>
> For more Docker CLI commands, refer to `https://docs.docker.com/engine/reference/run/`.
>
> Alternatively, check the frequently used commands from their official one-page cheat sheet here:
>
> `https://docs.docker.com/get-started/docker_cheatsheet.pdf`.

We will learn how to create our own custom Docker image and work with it. But before that, we need to learn how to manage package dependencies with the `requirements.txt` file in a Django project.

Working with the requirements.txt file

Different projects can have different dependency needs to run. That's why we use the `requirements.txt` file to capture all the third-party packages installed and used in a project. But as projects grow, we can realize that some of the dependency packages are useful only while we are developing locally, some are useful in our CI pipeline, and some are used only for production. Hence it is important that we break our `requirements.txt` file into multiple parts.

We need to split our `requirements.txt` file into multiple files that would be used for different purposes. This is common in large projects and is always better to implement as soon as we create a new project. This ensures that only the actually used packages are shipped to production.

Here is a structure seen in most large projects:

- `requirements-base.txt` – This file contains all the common third-party packages that are used in all the environments. For example, Django and Celery are third-party packages that are used across all the environments, so we should add them to the base file.

- `requirements-local.txt` – This file contains all the packages that are going to be used in the base file and are needed to run the project locally. This can contain a lot of mocks and also testing packages that developers use daily.

- `requirements-ci.txt` – This file contains all the packages that are going to be used in the base file and are needed to run the project in the CI pipeline.

- `requirements-stage.txt` – This file contains all the packages similar to the production environment but can be used to test any package upgrades without affecting the production environment.

- `requirements-prod.txt` – This file contains all the packages that are going to be used in the base file and any special packages that are needed in production.

Here is an example of a `requirements-base.txt` file that has only base packages:

```
Django==5.0
djangorestframework==3.14.0
redis==5.0.1
django-cacheops==7.0.2
```

Now our production packages would be stored in the `requirements-prod.txt` file, which has all the base packages along with the extra packages:

```
-r requirements-base.txt
gunicorn==21.0.1
```

The first line, `-r requirements-base.txt`, is important because it tells Python to include all the packages from the `requirements-base.txt` file and then add the packages mentioned in the `requirements-prod.txt` file. *Figure 11.5* shows the `requirements.txt` concepts:

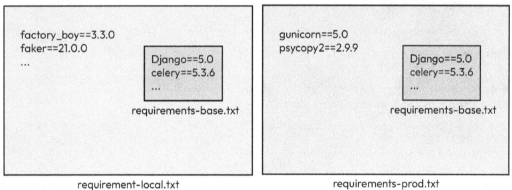

Figure 11.5: requirement.txt

There are other dependency tools in Python that can be used for package management. For example, there are `Poetry`, `pdm`, `pipenv`, but we will stick to `pip` and `requirements.txt` since we are using Docker and it will simplify our setup.

Now let us create our own Docker image for our Django project code that we can run locally.

Creating a Dockerfile for a Django project

A `Dockerfile` is a text document that has all the instructions needed to create a Docker image. We can consider a `Dockerfile` as a recipe that we can follow to cook any item. Let us take the boilerplate `Dockerfile` example we would need for our Django project. Create a `/myblog/backend/Dockerfile` file with the following content:

```
# Pull the official base image
FROM python:3.11
# Set environment variables
ENV PYTHONDONTWRITEBYTECODE 1
ENV PYTHONUNBUFFERED 1
# Install different Linux packages.
RUN apt-get update \
&& apt-get install gcc postgresql postgresql-contrib libpq-dev
python3-dev netcat-traditional -y \
&& apt-get clean
# set working directory and install packages using pip.
WORKDIR /app
COPY ./requirements ./requirements
RUN pip install -r requirements/requirements-local.txt

# Copy the Application code
COPY . /app/

COPY ./entrypoint.sh /app/entrypoint.sh

# Give permission to entrypoint.sh file as executable.
RUN chmod +x /app/entrypoint.sh

# run entrypoint.sh
ENTRYPOINT ["/app/entrypoint.sh"]
```

Save the content of the file as `Dockerfile` in `/<path to project>/<Django manage.py folder>/Dockerfile`. Let us learn what `Dockerfile` does:

- `FROM python:3.11` – We are downloading the base Python image where we shall install all the packages. One popular idea developers have with Docker is to use the `Alpine` image, but I strongly advise you to stay away from the `Alpine` image because the `Alpine` image with Python leads to longer build time, performance issues, and obscure bugs in production. (Read more about the issues with the `Alpine` Docker image at `https://pythonspeed.com/articles/alpine-docker-python/`.) I would strongly recommend using the default Python image if you are starting and don't have any specific requirements and space constraints.

- **Set Python environment variables** – Setting PYTHONDONTWRITEBYTECODE to 1 tells Python not to create the __pycache__ folders. Setting PYTHONUNBUFFERED to 1 tells Python to pass the stdout and stderr streams to be unbuffered. This helps to get an immediate output of the logs, and the Django management command output is instantly visible.

- **Install Linux packages** – We use different drivers and packages to connect to databases and other services.

- **Install Python packages** – We have first set the working directory, then copied all the requirements files. Run the pip command to install all the Python packages.

- **Copy application code** – Copy the application code and give execution permission to the entrypoint.sh file.

- Run the entrypoint.sh file.

Once our Docker container is initiated, we need to execute commands inside our container. For example, in a Django project, we want to run our migration command every time we start our Django server. This way, we can ensure that our database has applied all the latest database migrations. We can perform such operations using the entrypoint.sh script. We need to create an entrypoint.sh file. Here is an example of our entrypoint.sh file:

```sh
#!/bin/sh
if [ "$DB_HOSTNAME" = "postgres" ]
then
    echo "Waiting for Postgres..."
    while ! nc -z $DB_HOSTNAME $DB_PORT; do
      sleep 0.1
    done
    echo "PostgreSQL started"
fi
python manage.py migrate
exec "$@"
```

In the preceding snippet of code, we are first waiting for the Postgres database to be ready for connection. Once the database is ready to accept the connection, we will run the Django migration commands. The last line, exec "$@", is a very important command; it takes the command from the docker-compose.yaml file and replaces the parent process, rather than creating a different process. Read more about how the entrypoint.sh script works and what it can do here: https://stackoverflow.com/questions/32255814/.

Now run the following command to build a Docker image for your Django project:

```
> docker build -f ./Dockerfile -t django-in-production .
```

Figure 11.6 shows the output of the docker build command. In this step, we are downloading the Python image and then installing all our dependency Python packages.

```
docker build -f ./Dockerfile -t django-in-production .

[+] Building 10.0s (12/12) FINISHED
 => [internal] load build definition from Dockerfile                                                    0.0s
 => => transferring dockerfile: 37B                                                                     0.0s
 => [internal] load .dockerignore                                                                       0.0s
 => => transferring context: 2B                                                                         0.0s
 => [internal] load metadata for docker.io/library/python:3.11                                          1.0s
 => [internal] load build context                                                                       0.0s
 => => transferring context: 3.13kB                                                                     0.0s
 => [1/7] FROM docker.io/library/python:3.11@sha256:c0c5e12cd9fe77a556dea3bc71a71e16bb2fcb35974ce82215095d4cd279fb93    0.0s
 => CACHED [2/7] RUN apt-get update && apt-get install gcc postgresql postgresql-contrib libpq-dev python3-dev -y && apt-get clean    0.0s
 => CACHED [3/7] WORKDIR /app                                                                            0.0s
 => [4/7] COPY ./requirements/ /app/                                                                     0.0s
 => [5/7] RUN pip install -r requirements-local.txt                                                     7.9s
 => [6/7] COPY . /app/                                                                                   0.0s
 => [7/7] RUN chmod +x /app/entrypoint.sh                                                                0.3s
 => exporting to image                                                                                   0.6s
 => => exporting layers                                                                                  0.6s
 => => writing image sha256:cd67bffe57c82bb92b6bec82d328dfa2f3b0a78cc1df1af9f61afaf63adf08c1            0.0s
 => => naming to docker.io/library/django-in-production                                                 0.0s

Use 'docker scan' to run Snyk tests against images to find vulnerabilities and learn how to fix them
```

Figure 11.6: Output of the docker build command

To start the Docker container, run the following command:

```
> docker run -p 0.0.0.0:8000:8000 django-in-production python manage.
py runserver 0.0.0.0:8000
```

Figure 11.7 shows the output of the docker run command, where we are passing the python manage.py runserver 0.0.0.0:8000 command to the container and also exposing port 8000 to the host machine.

```
docker run -p 0.0.0.0:8000:8000 django-in-production python manage.py runserver 0.0.0.0:8000

Operations to perform:
  Apply all migrations: admin, auth, author, authtoken, blog, contenttypes, custom_user, sessions
Running migrations:
  No migrations to apply.
Watching for file changes with StatReloader
Performing system checks...

System check identified no issues (0 silenced).
December 17, 2023 - 11:14:11
Django version 5.0, using settings 'config.settings'
Starting development server at http://0.0.0.0:8000/
Quit the server with CONTROL-C.
```

Figure 11.7: Output of the docker run command

Now we can see that our Django application is running and is accessible via our browser. We can verify it by opening http://0.0.0.0.8000/admin/ in our browser.

We have created our Django Docker image, but we were still using a remote service for Postgres and Redis. Now we shall learn how to orchestrate different services such as Postgres and Redis together using docker-compose and connect to our Django project.

Composing services using docker-compose.yaml

Now let us explore how to create a `docker-compose.yaml` file that orchestrates our Django project, Celery, Celery Beat, Postgres, and Redis services together. Add the following code to the `/myblog/docker-compose.yml` file:

```
version: "3.8"
services:
  postgresql_db:
    image: postgres:16.1
    volumes:
      - ~/volumes/proj/dip/postgres/:/var/lib/postgresql/data/
    ports:
      - 5432:5432
    env_file:
      - ./.env
    environment:
      - POSTGRES_USER=${DB_USERNAME}
      - POSTGRES_PASSWORD=${DB_PASSWORD}
      - POSTGRES_DB=${DB_NAME}
  redis_db:
    image: redis:7.2
    container_name: redis_db
    ports:
      - 6379:6379
    command: redis-server --requirepass ${REDIS_PASSWORD}
    env_file:
      - ./.env
    volumes:
      - $PWD/redis-data:/var/lib/redis
      - $PWD/redis.conf:/usr/local/etc/redis/redis.conf
  blog_app:
    build:
     context: ./backend
     dockerfile: Dockerfile
    command: python manage.py runserver 0.0.0.0:8000
    volumes:
      - ./backend/:/app/
    ports:
      - 8000:8000
    env_file:
      - ./.env
    depends_on:
      - postgresql_db
```

```
        - redis_db

  celery:
    build: ./backend
    command: celery --app=config worker -l info
    volumes:
      - ./backend/:/app/
    env_file:
      - ./.env
    depends_on:
      - postgresql_db
      - redis_db
  celery_beat:
    build: ./backend
    command: celery --app=config beat -l info -S redbeat.
RedBeatScheduler
    volumes:
      - ./backend/:/app/
    env_file:
      - ./.env
    depends_on:
      - postgresql_db
      - redis_db
```

Let us learn what each section of the `service` configuration does (please note that the service names can be any other name as we are naming them at our convenience):

- `postgres_db` – Runs the Postgres database server. We are using a `.env` file to load all our environment variables values and then mapping the values in the `environment` section.

- `redis_db` – Runs the Redis database server used for caching and maintaining the queue for our Celery asynchronous tasks.

- `blog-app` – Runs the Django application.

- `celery` – Runs the Celery service to run asynchronous tasks.

- `celery-beat` – Runs the Celery Beat service to run periodic tasks executable.

Important note

We are using the same `Dockerfile` we created in the *Creating a Dockerfile for a Django project* section for our Django, Celery, and Celery Beat services. The `Dockerfile` and `entrypoint.sh` steps are the same; the only difference is the command we pass to the services.

One more important step: while running Docker in our local development setup, we have to give executable permission to the `entrypoint.sh` file in our host filesystem. We override the files in the Docker container using the `volumes` config, which mounts all the files into the Docker container. We use `volumes` to enable hot-reloading for Django development. By running the following command on Mac or Linux, in the `<project>/myblog/` folder, we can give appropriate permission to the `entrypoint.sh` file:

```
> chmod +x backend/entrypoint.sh
```

If we do not provide the appropriate permission to the `entrypoint.sh` file, then we will see the following error when trying to run the `docker-compose` command in the *Starting Django applications using Docker* section. In *Figure 11.8*, we can see the `exec: "/app/entrypoint.sh": permission denied: unknown` error. So, it is important that we add the appropriate permission.

```
docker compose up --build
 => [internal] load build context                                                                                    0.0s
 => => transferring context: 3.01kB                                                                                  0.0s
 => CACHED [2/8] RUN apt-get update && apt-get install gcc postgresql postgresql-contrib libpq-dev python3-dev -y && apt-get clean   0.0s
 => CACHED [3/8] WORKDIR /app                                                                                         0.0s
 => CACHED [4/8] COPY ./requirements ./requirements                                                                  0.0s
 => CACHED [5/8] RUN pip install -r requirements/requirements-local.txt                                              0.0s
 => CACHED [6/8] COPY . /app/                                                                                         0.0s
 => CACHED [7/8] COPY ./entrypoint.sh /app/entrypoint.sh                                                             0.0s
 => CACHED [8/8] RUN chmod +x /app/entrypoint.sh                                                                     0.0s
 => exporting to image                                                                                               0.0s
 => => exporting layers                                                                                              0.0s
 => => writing image sha256:9746a558b7d89d27f2ed7747e9f1e0a653cbe60db07fe704df268a9eea34ac18                         0.0s
 => => naming to docker.io/library/myblog-blog-app                                                                   0.0s
WARN[0001] Found orphan containers ([myblog-celery-beat-1 myblog-celery-1]) for this project. If you removed or renamed this service in y
our compose file, you can run this command with the --remove-orphans flag to clean it up.
[+] Running 2/0
 ⊓ Container myblog-postgresql_db-1  Running                                                                          0.0s
 ⊓ Container redis_db                Running                                                                          0.0s
Attaching to myblog-blog-app-1, myblog-postgresql_db-1, redis_db
Error response from daemon: failed to create shim task: OCI runtime create failed: runc create failed: unable to start container process:
 exec: "/app/entrypoint.sh": permission denied: unknown
```

Figure 11.8: Error due to a bad permission for entrypoint.sh

For a Windows-specific command to update the permission of the `entrypoint.sh` file, go to `https://www.educative.io/answers/what-is-chmod-in-windows`.

Now we have created the `docker-compose` file that has all the services needed for our Django application to run, but the configurations for each service need to be passed. For that, we create a `.env` file that has all the required variables.

Creating a .env file

We will create a `.env` file that has all configuration values and also the sensitive secret values needed by different services. Here is an example of the environment variables that our `docker-compose` would need to run all the different services. Create a `/myblog/.env` file and add the following code:

```
DEBUG=True
DJANGO_ALLOWED_HOSTS=*
DB_ENGINE=django.db.backends.postgresql
DB_NAME=django_in_production
```

```
DB_USERNAME=root
DB_PASSWORD=root
DB_HOSTNAME=postgresql_db
DB_PORT=5432
REDIS_HOST=redis_db
REDIS_PORT=6379
REDIS_PASSWORD=redisPassWord
```

Now we have defined all the configuration values in the .env file and saved the .env file in the same directory where our docker-compose.yaml file is located. *These values can be different and can be updated as per your preference.* As a next step, we have to load the .env file for all the services in the location where we want them to be accessible. Taking the example of our docker-compose. yaml file, we are loading the .env values as follows:

```
version: "3.8"
services:
  blog-app:
    ... other settings
    env_file:
      - ./.env
    ... other settings
  celery:
    ... other settings
    env_file:
      - ./.env
    ... other settings
```

Our blog-app service and celery service would have the environment variables loaded from the .env file.

There are other ways to pass environment variables in Docker. You can read more here: https://docs. docker.com/compose/environment-variables/set-environment-variables/.

The environment variables that are passed need to be accessed in Django and Celery services. Let us learn how we can access environment variable values using Python in our backend services.

Accessing environment variables in Django

Once we add the environment variables, we need to access them in our Django application running inside the Docker container. The environment variables can be accessed using Python's os library. For example, let us access the DEBUG flag set in our .env file:

```
import os
DEBUG = os.environ.get("DEBUG", False)
```

Similarly, we can access other environment variables and make our `settings.py` file dynamic based on the values set in the `.env` file. We can configure the Django settings with the environment variables picked from the `.env` file so that the secret values don't get added to the version control.

> **Important note**
>
> Please remember to add the `.env` file to the `.gitignore` file. We do not want the `.env` file or any sensitive value that can cause any security vulnerability to be committed to the Git history.

There is another package available, `django-environ` (https://github.com/joke2k/django-environ), which can load the environment variables in a better format. But I will leave it to you to read and decide whether to use the basic `os` library or `django-environ`. As a personal preference, I like to use the basic `os` library, since using a third-party package always adds an overhead of maintenance. Reading simple environment variables shouldn't require an additional package.

Now let us start our Django application using Docker Compose.

Starting a Django application using Docker

Our Django project is now set up with Docker and we can start our project by executing the following command:

```
> docker-compose -f docker-compose.yaml up --build
```

Now Docker will download the images and build all the five different services mentioned in the `docker-compose.yaml` file. Our Django application is now running independently on our local system without depending on any of the remote services.

To verify that all the services are running properly, we can use the following command:

```
> docker ps
```

Figure 11.9 shows the result of the `docker ps` command and all the five containers running in our local system:

```
docker ps
CONTAINER ID  IMAGE                COMMAND               CREATED            STATUS            PORTS                     NAMES
eaef80de3c5b  myblog-celery-beat   "/app/entrypoint.sh …"  About a minute ago  Up About a minute                           myblog-celery-beat-1
dbed2912cd42  myblog-celery        "/app/entrypoint.sh …"  About a minute ago  Up About a minute                           myblog-celery-1
50667576a3cb  myblog-blog-app      "/app/entrypoint.sh …"  About a minute ago  Up About a minute  0.0.0.0:8000->8000/tcp    myblog-blog-app-1
e070bc17a51c  postgres:15          "docker-entrypoint.s…"  2 hours ago        Up About a minute  0.0.0.0:5432->5432/tcp    myblog-postgresql_db-1
81bd35bbd3f0  redis:7.2            "docker-entrypoint.s…"  2 hours ago        Up About a minute  0.0.0.0:6379->6379/tcp    redis_db
```

Figure 11.9: The output of the docker ps command

Docker is a powerful tool that helps developers to independently run services on their local system. We have utilized the power of Docker to successfully run our Django application locally using different Docker containers.

Summary

In this chapter, we had a high-level overview of how to set up a Django application using Docker. We took a boilerplate template that had Postgres, Redis, Django, Celery, and Celery Beat to run the service locally using Docker. We have not deep-dived into the topics because Docker itself is a vast topic and it is expected that you have some understanding of Docker so that you can use this chapter as a reference to set up your project.

In the next chapter, we shall focus on how we can implement a CI pipeline using GitHub Actions, setting up Git hooks, and using version control with different branching strategies. We shall also discuss a few points on how to set up a code review process.

12

Working with Git and CI Pipelines Using Django

In *Chapter 11*, we learned how to integrate Docker into a Django project. Docker helps us to run our Django project on any platform seamlessly. In this chapter, we shall explore how to use Git efficiently while working with a Django project. We shall learn about the best practices for using Git and how to integrate Git hooks into the development cycle to increase the productivity of developers.

We will be learning about the following topics in this chapter:

- Using Git efficiently
- Working with GitHub and setting up GitHub Actions
- Setting up PR review and code guidelines

Technical requirements

In this chapter, we shall work with advanced Git features and integrate GitHub Actions into our project. We are expecting our readers to have a good understanding of Git and some knowledge about **continuous integration** (**CI**) pipelines using GitHub Actions.

If you are not familiar with Git, we highly recommend you learn the basics of Git and come back to this chapter. Here are a few resources, including books and free courses, that you can follow:

- *Git for Programmers* by Jesse Liberty
- *Mastering Git* by Jakub Narębski, Jakub S Narebski
- *Version Control with Git* by Udacity (`https://www.udacity.com/course/version-control-with-git--ud123`)
- *Pro Git* by Scott Chacon and Ben Straub (`https://git-scm.com/book/en/v2`)
- *RY's Git Tutorial* by Ryan Hodson (`https://www.amazon.com/Rys-Git-Tutorial-Ryan-Hodson-ebook/dp/B00QFIA5OC`)

Here is the GitHub repository that has all the code and instructions for this chapter: `https://github.com/PacktPublishing/Django-in-Production/tree/main/Chapter12`

Using Git efficiently

Version control is one of the most important tools in the **software development lifecycle** (**SDLC**). We will not go into the basics of Git in this chapter, but rather we will discuss the good practices one should follow while using Git. Let us look into a few good practices that are recommended while using Git:

- Always create a new branch event for small changes. A lot of times, developers get tempted to directly commit their changes to `main`, `master`, `develop`, or any other name used for the default branch, depending upon what they are using.

- Add a `.gitignore` file as soon as you create a project.

- Never allow anyone to have access to push code to the default branch. Every piece of code should be merged to the default branch using a **pull request** (**PR**).

- Always make sure Git is configured locally and has the developer's name, email, and other relevant details.

- Use the `git merge` command strategy while merging two branches and avoid using rebase and rewriting of history, unless you know what you are doing. We will discuss this in detail in the following section.

- Always use meaningful and verbose commit messages. Create a commit message guideline and make sure developers follow the commit message format.

- Always use `git revert` to perform and rollback of code. Never change the published history of your code base.

- Use Git hooks as much as possible to automate your workflow.

- Use Git tags to release new code in a project. We shall discuss in detail how we should utilize Git tags in the later part of the chapter.

- Decide on a **Git workflow** and make sure the developers stick to it.

> **Please note**
>
> The previously mentioned best practices on Git are opinionated and are situation based. Readers can always choose a different workflow that will fit/solve their problem statement. I am not able to follow all the previously mentioned recommendations while working on different projects. Hence it is always possible to reason out a different workflow as opposed to what is mentioned.

Now, let us start deep-diving into a few topics that can help you and your team work efficiently.

Branching strategy for Git

Developers should always create a new branch while working on a code base. While working with branches, there are a few considerations or conventions developers should follow. In this section, we shall discuss all the conventions a software developer should consider to improve their development process using Git:

- **Branch name**: It is important to decide on a branch naming convention. A branch name should provide just enough information that can help anyone easily identify what the changes are in the branch. There are different types of conventions different teams and companies follow, but the most important thing is that everyone should follow the same convention in the team. Here are a few examples that are used in the industry; feel free to discuss and choose a branch convention that suits your team:

    I.   `<branch type>/<ticket or issue number>/<branch name>`: Here, we start the branch name with the kind of work it is addressing, such as `feature`, `bugfix`, and `improvement`. Then we would add the issue number or the JIRA ticket number followed by a standard branch name. For example, if we are doing a feature and our JIRA number is DJ-112, which would add a comment feature to the blog, then our branch name should look like `feature/DJ-112/integrating-comments`.

    II. `<developer or team name/<branch type>/<branch name>`: Here, we would start the branch name with the team or developer name who is working on the branch, followed by the type of work, such as `feature` or `bugfix`. For example, if I am working on adding a comment feature then the branch name would look something like this: `argo/feature/integrating-comments`.

    Developers can discuss and decide on any other naming convention that fits well within the team, but it is important to choose a particular pattern and only allow the following pattern to be used while working in the codebase. This can be enforced using local Git hooks and server hooks.

- **Using protected branches**: There should be protected branches in your project where no one can directly push any code. The only way any code can go to these branches should be via PRs. These branches should be used for production deployment or creating tags that are used to deploy to production.

- **Using a branching workflow**: There are numerous strategies developed by different companies that explain the branching strategy. You can read through different famous branching strategies and decide on which branching strategy to follow. Here are a few articles that you can explore:

    I.   Here is an article from the Git official website: `https://git-scm.com/book/en/v2/Git-Branching-Branching-Workflows`.

    II. Here is the link to one of the earliest Git workflows by *Vincent Driessen* that is still referred to across organizations: `https://nvie.com/posts/a-successful-git-branching-model/`.

III.   Here is the GitHub workflow: `https://docs.github.com/en/get-started/quickstart/github-flow`.

IV.   Here is the GitLab workflow: `https://docs.gitlab.com/ee/topics/gitlab_flow.html`.

Once we have decided on our branch workflow, we need to focus on the most important Git command, `git commit`.

Following good practices while using git commit

Once our code changes are done, we will use a `git commit` command. Every `git commit` command can be considered as a log book that can be used to identify code changes made in the past. Hence, we must write proper commit messages for every commit so that our Git history is relevant and it is easy to identify code changes.

Commit messages are one of the most important elements in the SDLC. They are used to identify any change made to the code. Developers must write appropriate commit messages for every commit they make to create maintainable code. Let us discuss a few standard commit message formats that are widely used in the industry.

Conventional Commits

The Conventional Commit documentation states the following:

The Conventional Commits specification is a lightweight convention on top of commit messages. It provides an easy set of rules for creating an explicit commit history; which makes it easier to write automated tools on top of.

The Conventional Commits guidelines suggest that the commit message should be structured as follows:

```
<type>[optional scope]: <description>
[optional body]
[optional footer(s)]
```

This helps to easily identify what type of work the code change reflects, and you can also write automated tools and workflows based on the commit messages. The `<type>` mentioned in the commit message helps in identifying what type of change the particular commit is introducing. `scope` defines the scope of the change. And finally, in the `description` block, we add a short description of the overall change. The conventional commit documentation provides an appropriate example to explain different types of commit messages. For further details please visit `https://www.conventionalcommits.org/`.

Use JIRA and GitHub Issues in commit messages

Project management tools are used to track the progress of the development tool. For example, all open source projects hosted in GitHub use GitHub Issues to track any feature request/bug. When we add GitHub issue numbers to our commit messages, GitHub then automatically links the code commit and the issue. This helps in aligning the team to actively track any progress for a given task. Similarly, revisiting the code commit history helps to get context on the what, why, and how of any code change.

If your team uses JIRA or some other tool for project management, then adding the JIRA ID to the commit message helps in identifying and tagging issues faster and better. There are different formats used in the industry and it is best to discuss within the team to finalize the commit message format that suits your team well. Let us look at one of the common formats used:

```
<type> [<JIRA ID>]: <Short description>
```

> **Read more**
>
> For more details, please visit the official documentation of GitHub on how we can use GitHub Issues efficiently: `https://docs.github.com/en/issues/tracking-your-work-with-issues/about-issues`. To integrate JIRA or other project management tools using GitHub, visit `https://github.blog/2019-10-14-introducing-autolink-references/`.

As the team grows and the project evolves, it is important to add automated tools that can enforce commit message formats. There are a few tools that can help you enforce a good commit message structure. Readers are advised to go through the tool's official documentation and understand the tools that suit them better:

- `commitlint` (`https://github.com/conventional-changelog/commitlint`)
- `gitlint` (`https://github.com/jorisroovers/gitlint`)
- `commitizen` (`https://github.com/commitizen-tools/commitizen`)
- `cz-cli` (`https://github.com/commitizen/cz-cli`)

Now, let us learn about a few tools that can help developers work with Git with ease.

Tools with Git

There are plenty of tools available in the market that can help us work better with version control. Let us look into the top tools that one can explore:

- **git-sim** (`https://github.com/initialcommit-com/git-sim`): When you are a beginner, it can be daunting to understand what different commands are doing in Git. git-sim is a command-line visualizer that can help beginners understand what basic commands are doing.

- **GitHub Desktop**: GitHub Desktop is a GUI for Git. It is platform agnostic and can be used for different code hosting platforms such as GitLab and Bitbucket.

- **GitKraken**: GitKraken is one of the oldest Git GUIs and also one of the most popular GUI tools available in the market.

We have learned how to create a new branch, and then make commits. We saw the good practices on how the branch name should look and what would be a good commit message, but how can we ensure that these guidelines are followed by the developers? That is where the concept of Git hooks comes into the picture.

Integrating Git hooks into a Django project

Git has a way to execute custom scripts whenever a certain important action occurs in Git; these custom event triggers are known as Git hooks. Git provides two types of hooks: client-side hooks run on the developer machine and events such as `commit`, `merge`, and `push`, while server-side hooks run on the remote server and events such as `pre-receive` and `post-receive`. Git hooks reside in the `.git/hooks` folder. One can easily create the Git hooks and integrate them, and different frameworks help us use them. We shall not get into the details of how to add a custom hook manually but rather work with the `lefthook` framework. Though one of the most popular Git hook frameworks for the Python ecosystem is `pre-commit`, we shall not work with `pre-commit`. Rather, we shall learn how to use `lefthook` since it is faster and more flexible.

Using lefthook

Lefthook is a Git hook manager written in Go. Since it is a dependency-free binary, it is easy to install and run in any environment. It can run commands in parallel, which makes it 3x faster than `pre-commit` and faster than most of the Git hook managers available. Lefthook documentation is evolving, but the community is quite active and supportive of resolving any blockers. For more details, check the project repository at `https://github.com/evilmartians/lefthook`.

Now, let us look into how to integrate `lefthook` into our Django project. The official documentation has all the mentioned details on how to install it: `https://github.com/evilmartians/lefthook/blob/HEAD/docs/install.md`.

For Mac systems, we can use the following command, using `homebrew`, to install lefthook:

```
brew install lefthook
```

Linux users can install lefthook via the APT package manager:

```
curl -1sLf 'https://dl.cloudsmith.io/public/evilmartians/lefthook/
setup.deb.sh' | sudo -E bash
sudo apt install lefthook
```

For Windows users, lefthook can be installed by using scoop (`https://scoop.sh/`):

```
scoop install lefthook
```

Alternatively, windows users can use winget to install lefthook (`https://github.com/microsoft/winget-cli`):

```
winget install evilmartians.lefthook
```

Now, let us integrate lefthook into our project. To learn lefthook, we shall create a simple project:

1. Create a `demo-project` folder and add a new `demo-project/demo_script.py` file with the following code:

    ```
    def hello_world():
        print("hello world")
    hello_world()
    ```

2. Add `git` to the folder by using the following command:

    ```
    git init
    ```

3. Now, we shall add `lefthook` to our project by using the following command:

    ```
    lefthook install
    ```

    This command will add a new `lefthook.yaml` file to the project folder that will have a bunch of sample commands that are commented.

4. Let us add a sample Git hook that would check the configuration in the generated `lefthook.yaml` file that has sample checks:

    ```
    pre-commit:
      commands:
        demo-echo:
          run: echo "Just Demo command"
        demo-echo2:
          run: echo "Just Demo command 2"
    ```

5. The `pre-commit` commands run whenever a developer would make a commit. So, when we perform the `git commit` command, we see Lefthook run the following custom hooks:

    ```
    > git commit -am "Init commit"

     lefthook v1.5.5  hook: pre-commit
    |  demo-echo  ❯
    Just Demo command
    |  demo-echo2  ❯
    ```

```
Just Demo command 2
_____

summary: (done in 0.03 seconds)
✔   demo-echo
✔   demo-echo2
[main 30019a4] Init commit
 2 files changed, 9 insertions(+)
create mode 100644 demo_script.py
```

The echo command that is used can be easily replaced by any other command that we want to execute.

Lefthook is versatile in nature and can be used to add Git hooks in different programming languages such as Python, Javascript, and Shell Script.

> **Read more**
>
> For more details, visit the official documentation of lefthook that explains all the features it provides: https://github.com/evilmartians/lefthook/blob/master/docs/usage.md.

The beauty of lefthook is that it is language agnostic, so we can use lefthook to create Git hooks for different languages. Here are a few of the lefthook scripts I generally configure:

- Linting of all the staged files using Black and Flake8
- Enforce the branch name as per the convention
- Enforce the commit message format as per the convention
- Stop any large file to be committed to the repository

Lefthook helps in client-side Git hooks. The other type of Git hook is the server-side hook. Unfortunately, GitHub does not support server-side hooks. But such custom logic can be implemented using GitHub Actions and whenever a developer is not following the guideline, we shall fail the GitHub Actions pipeline.

Now that we have learned how to force the developers to adhere to the guidelines, let us focus on how we can merge our code changes and release the code.

Using git merge versus git rebase

git merge and git rebase have their advantages and disadvantages. In *Figure 12.1*, we can see how the history of commits gets affected when we use the git merge or git rebase command. So, depending upon the use case, developers should choose whether to select git merge or git rebase. But leaving that option to the developers can also be dangerous, especially if your team has many junior developers. Hence, I generally prefer to choose git merge as the default option whenever creating a process in the team since it is easy to understand, so everyone has very little chance of messing up.

The advantages of using `git merge` are the following:

- It preserves the full repo history
- It makes it easy to resolve conflicts between different branches
- It makes it easy to revert any code change

git merge versus git rebase

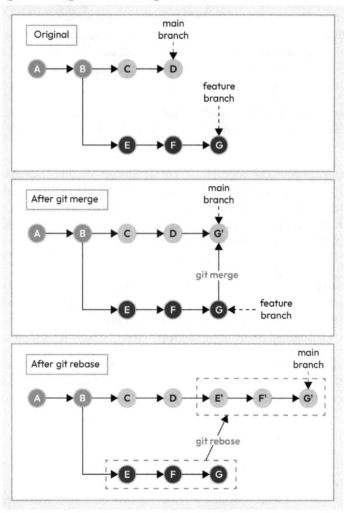

Figure 12.1: Image illustrating git merge versus git rebase

GitHub and other similar platforms give another option to choose from while merging using the UI (i.e., **Create a merge commit** or **Squash and merge**). Both have their advantages and disadvantages, and it depends on the situation. The biggest advantage of selecting **Squash and merge** is that it keeps the commit history clean, but the disadvantage is that you lose out on all the history of the commits made. While the **Create a merge commit** option would simply add a new merge commit and you would have the past commit history intact, I prefer using the **Create a merge commit** option for most of my use cases.

Once the code is merged, we are ready to deploy our code and release our feature to the users.

Performing code release

Shipping your code to production is the most fascinating experience for any developer. Watching one's code used by tens, thousands, or even millions of users is an absolute pleasure and joy to have. Hence the code release process must be defined well. Every team can decide on their code release process; one can have a manual deployment in each server or have a sophisticated **continuous deployment (CD)** process.

I generally prefer to build an automated one-click deployment process for services. In *Chapter 13*, we shall discuss more on how we can perform one-click deployment using GitHub Actions and AWS Elastic Beanstalk.

One of the most important points for code release is using Git tags. We should always use Git tags for each code release so that we can identify the exact code that is being executed on the servers. Each release can be labeled with a version number, and there are different versioning strategies available in the industry, such as semantic versioning (`<major>.<minor>.<patch>`), build number, and calendar versioning. The most common type of versioning is semantic versioning and most open source projects use semantic versioning to track their releases.

Another important aspect of code release is a **changelog** or **release note**. Changelog or release notes help us to identify all the changes that are going to the production. If we maintain a good commit message and track all the PRs that are merged between the previous release and the new release, then we can easily collate a changelog. Numerous tools help in creating automated release notes.

No developer would intentionally deploy buggy code to production, but somehow, bugs will crawl their way to production. So, it is important to monitor our system for any bugs. We shall learn more about how to perform application monitoring in *Chapter 14*. Now let us learn how to fix our code.

Performing hot-fixing on code

The first solution to fix any issue in production is **revert** A developer should always revert their change first and then think of fixing the bug. It can be a single-character change that can fix the code, and the developer might want to make small code changes to fix the code but cannot guarantee the stability of the system. Hence, it is always recommended to revert any code change made to the system and then try to find the issue and solve it. The first thing a developer should always prioritize is the stability of the system and avoiding impacting any of the end users of the product.

However, it is not always possible to revert code changes. In these cases, the next best thing to do is create a branch from the released Git tag and add the code change to that branch. Never deploy any other code change apart from the code fix that needs to go.

If a team follows the preceding steps, then any code release will always go smoothly, because the team prioritizes the stability of the product and user experience over anything else.

Now, let us explore how to work with cloud platforms such as GitHub and GitLab.

Working with GitHub and GitHub Actions

Multiple platforms such as GitHub, GitLab, and Bitbucket can provide code repository hosting solutions. GitLab is an open source project that can be also used to host our own GitLab server. Currently, GitHub and GitLab are the main market leaders. We shall be discussing GitHub in this book and will not be looking into GitLab; if you are using GitLab or any other platform, you can skip this section and move to the next section of the chapter.

Here are a few important features/settings/configurations we would like our readers to be aware of while working with GitHub:

- **Always use Organization for projects that are not personal**: GitHub provides an interface to create an organization, which will help in organizing all the repositories better.

- **Enable 2FA for all users**: A code base is one of the most sensitive components of an organization hence security measures should be the priority.

- **Configure rulesets**: Branch rulesets help to enforce guidelines on how to use enforce rules for given branches/tags. As the team grows, you might not want everyone to have permission to push code to a particular branch.

- **Automatically delete head branches**: This deletes the branches automatically after they are merged.

- **GitHub Copilot**: GitHub Copilot is an AI tool that assists developers in writing code faster. The code logic and other key coding concepts should be taken care of by the developer, but Copilot can boost your productivity. I have been using Copilot for writing test cases and other low-level code and it has helped me to write those features faster.

- **Use GitHub code owners**: GitHub code owners can help in structuring the ownership of code in a better way. In a repository, different teams own different parts of the code, hence using code owners helps in defining the ownership and improves the code review process.

- **Use GitHub Teams**: GitHub provides teams features; for example, the frontend developers can be part of the frontend engineering team.

- **Use Dependabot**: Dependabot is an automated dependency update bot that checks for security fixes regularly in the repository. It can be considered the first line of defense to safeguard any potential vulnerability added to the project due to the packages used.

There are multiple other features that GitHub provides and each team has different needs and use cases for them. Readers are advised to use all the relevant features and automation steps that can help them to boost their productivity. Now let us look into how to use GitHub Actions to create our own CI pipeline.

Working with GitHub Actions for the CI pipeline

As developers, we constantly try to automate all repetitive tasks, streamline our workflow, and ensure the smooth deployment of our code. Tools such as Jenkins, CircleCI, Travis CI, and GitHub Actions help us achieve such automation workflow. GitHub Actions, which can be used for CI/CD pipelines, can help us automate workflows that boost the productivity of the development team. In this section, we shall learn how we can utilize GitHub Actions to automate our workflow for Django projects.

> **Read more**
>
> The quick start guide of GitHub Actions explains how to get started with GitHub Actions. We are expecting our readers to be familiar with the basics of GitHub Actions or follow the guide to come up to speed before moving to the next section: `https://docs.github.com/en/actions/quickstart`.

Now, let us learn how to set up our CI pipeline for our Django project.

Setting up a CI pipeline for Django using GitHub Actions

CI is considered to be one of the best practices for software development that helps developers merge their code to a shared branch, for example, the `main` branch, and then the automation steps start running automation tests and builds. It is often misunderstood that a CI pipeline is helpful only when we want to automate our builds and deploy, but in reality, CI pipelines help in improving developer productivity. For example, when a developer writes a feature and pushes their code to their branch, a good CI pipeline, automation unit, and integration tests can be run against their branch to identify any potential bug/issue that can be caught early on.

Now, let us learn how we can create the basic actions that run the unit tests for our Django project.

> **Important note**
>
> GitHub Actions will work only when the `.github` folder is present in the project parent folder. Hence, the code example in the official Git repository for this book will not work. Please follow the instructions mentioned in the code section to create your actions: `https://github.com/PacktPublishing/Django-in-Production/tree/main/Chapter12#setting-up-ci-pipeline-for-django-using-github-actions`.

Setting up GitHub Actions with an example

We are going to run the Django unit tests whenever someone pushes their code to GitHub.

Create a file in `.github/workflows/django-ci.yaml` with the following code. We are defining our actions and setting the right environment variables:

```
name: Django CI Test Cases
on: push # Run this Action on every code push made.
env:
  DB_NAME: blog_testdb
  DB_USERNAME: root
  DB_PASSWORD: root
  DB_HOSTNAME: 127.0.0.1
  DB_PORT: 5432
  DB_ENGINE: django.db.backends.postgresql
  REDIS_HOST: 127.0.0.1
  REDIS_PORT: 6379
  REDIS_PASSWORD: redisPassWord
```

The name of the workflow is defined using the `name` attribute; here we are naming our workflow `CI Django Test Cases`:

```
name: CI Django Test Cases
```

The on attribute defines when the workflow should be executed. Multiple events can trigger GitHub Actions; check the official documentation for more details: `https://docs.github.com/en/actions/using-workflows/events-that-trigger-workflows`:

```
on: push
```

The env section is the same as the environment variable. It is used to set any environment variable value for the code repository. We are setting the Postgres credentials using the `env` attributes here:

```
env:
  DB_NAME: blog_testdb
  ... # Other environment variables as needed
```

Now, let us add the GitHub jobs to our `.github/workflows/django-ci.yaml` config file. Please note that we are appending the code to the previous section:

```
jobs:
  # Label of the container job
  container-job:
    runs-on: ubuntu-latest
    services:
      postgres:
        image: postgres:latest
        env:
```

```
        POSTGRES_USER: root
        POSTGRES_PASSWORD: root
        POSTGRES_DB: blog_testdb
      ports:
        - 5432:5432
      # Set health checks to wait until Postgres has started
      options: >-
        --health-cmd pg_isready
        --health-interval 10s
        --health-timeout 5s
        --health-retries 5
    redis:
      image: redis:latest
      ports:
        - 6379:6379
```

Inside the `job` attribute section, we can define multiple jobs of a workflow. For example, if we want to run linting, unit tests, and integration tests inside the same workflow, then we can create different jobs. In our case, we have created `container-job` (note – it can be given any name), which runs the Django test.

The `runs-on` attribute defines the base layer on which the job executes. Unless there is a specific need or customization need, it is recommended to start with `ubuntu-latest`:

```
runs-on: ubuntu-latest
```

The `services` attribute defines all the external services that the job needs. For example, we need Postgres to run our unit test since we have a database connected to our Django application. Similarly, we can also have Redis as a separate service connected to this job. For more details, check the official documentation: `https://docs.github.com/en/actions/using-containerized-services/about-service-containers`.

Now let us add `steps` to our actions:

```
jobs:
  container-job:
    services:
      . . . # Service config
    steps:
      - uses: actions/checkout@v3
      - name: Set up Python 3.11
        uses: actions/setup-python@v3
        with:
          python-version: 3.11
      - name: Install Dependencies
```

```
    run: |
      cd backend
      python -m pip install --upgrade pip
      pip install -r requirements/requirements-local.txt
  - name: Run Django migrations
    run: |
      cd backend
      python manage.py migrate
  - name: Running Django Test cases
    run: |
      cd backend
      python manage.py test --no-input
```

Once we have set up the services, we will now define `steps`. In `steps`, we will group all the commands that need to be executed in the job:

```
steps:
    - uses: actions/checkout@v3
    - name: Set up Python 3.11
      uses: actions/setup-python@v3
      with:
        python-version: 3.11
```

First, we check out the code from the repository and make it available in the `uses: actions/checkout@v3` GitHub Actions job, then we set up the Python version for the code to be executed.

The next step is installing the dependency by using the `pip install -r requirements/ci-requirements.txt` command:

```
    - name: Install Dependencies
      run: |
        cd backend
        python -m pip install --upgrade pip
        pip install -r requirements/requirements-local.txt
```

The name attribute is used to name a particular step in a job; it is easy to identify a step using a name in the GitHub Actions UI.

```
    - name: Run Django migrations
      run: |
       cd backend
       python manage.py migrate
    - name: Running Django Test cases
```

```
run: |
  cd backend
  python manage.py test --no-input
```

In our custom actions, we have connected the Postgres database, and then run our Django test cases whenever someone makes a code commit and pushes their code to GitHub. In our example, we have performed a simple operation of running our Django test cases every time there is a code push made by the developer.

Testing out GitHub Actions

Now, we are ready to run our Django-related commands. First, we run migration on our Django project and then execute the Django test cases. If our test cases pass, then we will see a success in our GitHub Actions pipeline; otherwise, we would see the failed test cases.

Whenever we push code changes to our GitHub server, our pipeline will be automatically triggered. We can make a code commit and push our code with the following command:

```
git commit -am "Test pipeline"
git push
```

This will push the local changes to our remote GitHub repository and also trigger the pipeline. We can see *Figure 12.2* showing the workflow's successful run with the title same as our `git commit` message:

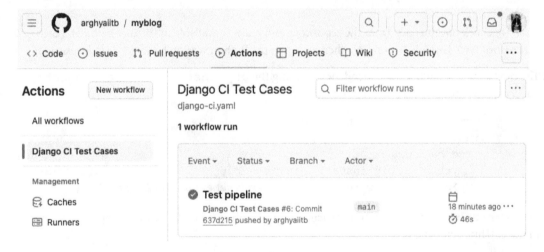

Figure 12.2: Showing GitHub Actions pipeline triggered on code push

When we go inside the workflow, we can see only one entry since we have just one job defined. For complicated projects, there can be multiple jobs. *Figure 12.3* shows the job running when we push our code:

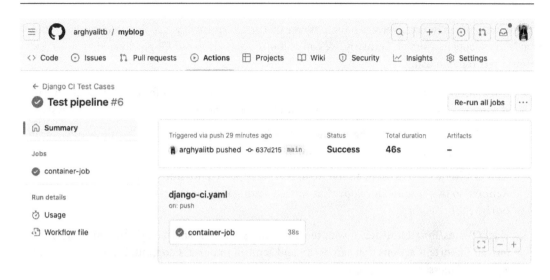

Figure 12.3: Showing a GitHub job

When we go inside `container-job` again, we can see all the steps we defined in our `django-ci.yaml` file. *Figure 12.4* shows all the steps that are running:

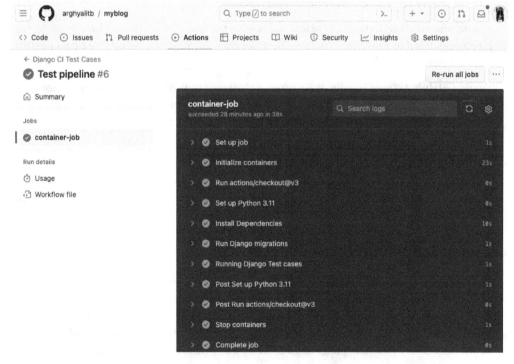

Figure 12.4: Showing all the steps we have configured in our django-ci.yaml file

GitHub Actions can be considered a remote server that can perform any task that a developer would be able to execute in a server. Similarly, we can add more actions to our CI pipeline. Here are a few actions I recommend using:

- **Code coverage threshold**: Adding a code coverage threshold test for every code push will ensure developers actively pursue writing unit tests that can ensure the stability of the system

- **Code linter**: Linting should be enabled so that if developers don't set up Git hooks and a local linter, then it can be caught in the GitHub pipeline

- **Run Git hooks**: It is always advisable to re-run all the client-side Git hooks using GitHub Actions, so that if some developer has missed setting it up or bypassed the checks then they can be easily caught

- **Slack/Teams/Email notifier**: If your team is using Slack or Teams, you should integrate Slack into all important actions that can notify and send all important communications directly as a personal message or in a group

Now, let us learn a few good practices while using GitHub Actions:

- **Always use cache**: GitHub Actions support caching of steps; using cache can improve the execution time of long-running repetitive steps such as `pip install`. For more details, visit `https://github.com/actions/cache`.

- **Create semantic versions**: We have learned how to use semantic versioning for our project; it is recommended to use GitHub Actions to make releases. Similarly, we can create automated release notes for each of our releases.

- **Verify GitHub Actions locally**: It becomes a challenge to verify GitHub Actions since each update in the YAML configuration will mean we have to run the pipeline again. Hence, a better way to verify the changes would be by using the project at `https://github.com/nektos/act`.

- **Upload artifact**: As the team matures, there will be actions that use the same build steps and artifacts, so rather than repeating the same build/creation process, it is recommended to use `actions/upload-artifact` to share artifacts between jobs. For more details, visit `https://github.com/actions/upload-artifact`.

Recommended GitHub Actions resources

The experience of using the CI pipeline in GitHub has improved ever since the introduction of GitHub Actions in 2018. Every week, developers create more open source projects for GitHub Actions and share them with different developers across the globe. Therefore, before building anything in GitHub Actions from scratch, I would recommend one go and search for the existing actions and see if they can solve your use case. I generally prefer to go to `https://github.com/sdras/awesome-`

`actions` and search for the relevant actions before custom building. However, one should also check if the action is maintained currently or not. I generally ask these questions before using any third-party action:

- How many users have used the actions? We can find this information on the GitHub project repository page.

- What are the active issues in the project? Is the community actively responding to GitHub Issues? Do the active issues have security vulnerabilities that the project maintainer is not addressing?

- When was the last time the project saw any code update?

- How many stars do the actions have?

The preceding questions can only be considered as an outline that can help the developer decide whether they should use the third-party GitHub Actions.

Now, let us check a few of the top actions that I use frequently, apart from the official GitHub Actions:

- Send slack message whenever there is some action performed on the code repository using the pipeline, such as deployment (`https://github.com/marketplace/actions/slack-send`)

- Perform linting using GitHub Actions (`https://github.com/marketplace/actions/super-linter`)

- Check for any secret key accidentally committed in the code repository (`https://github.com/marketplace/actions/trufflehog-oss`)

Once our feature is tested and we have all the pipelines passing, the next task for us is to perform a code review. Code review is a process that is pivotal for any development team. In the next section, we shall discuss how we can incorporate a good code review process and structure for a development team.

Setting up code review guidelines

Code review is the quality assurance process of any code change made to the code base. It is one of the loved and equally hated processes in the development cycle. The code review process, if implemented properly, can help create a great experience for the developers and fast track the product development cycle. But if the code review process is implemented appropriately, then it can become a bottleneck for the velocity of the company, and developers might become hostile toward doing code review, and a lot of times they may cause more issues. The code review process should be implemented by the early members of the team and should be continuously improved and adapted as the company/product/team grows. There are no definitive guidelines and it falls to team members to set up the guidelines that suit their situation. However, let us look into a few code review guidelines that are followed across the industry.

Context and description

Whenever a code change is sent for review, making sure that there is a good context and description added to the PR helps produce a better and faster review. We can add the following:

- **What?**: What does the code change intend to achieve? Explain the outcome we are trying to achieve via the code change. You should try to add all the context here and mention all the relevant non-technical documents (if applicable).

- **How?**: How is the solution implemented in the code? Give a high-level overview of how the code is implemented. It is also helpful to attach any architecture doc or design doc in this section.

- **Why?**: Why are we implementing the code changes (if applicable)?

Though in this book we are focusing primarily on backend changes, code review guidelines can be applicable across different coding teams. We should always try to attach all the relevant screenshots or videos linked to the code changes in the PR description.

Short code changes to review

One of the most crucial recipes to create a good code review process is to make sure the code changes are short. Setting a guideline to have a maximum of 300-500 lines of code to be reviewed in a PR helps in creating a good process.

But then how would a feature that needs more than 1,000-2,000 lines of changes be reviewed? For such features, it is always better to have a main feature branch and let developers create a smaller feature branch and merge them into the main feature branch. By using this approach, the code reviewer will always have a smaller code change to review and once the code changes are merged to the main feature branch, it is easier to glance at all the changes made to the humongous branch and review them.

Review when the code is ready

The code review should be done only when the PR is ready to be reviewed. We can use a few checklists that can help developers understand if their code is ready to be reviewed or not. Here are a few checklist items that I generally follow:

- Remove all the debug statements or print statements or any commented statements (these can be taken care of by the linter itself).

- Test the code changes in the development environment.

- Make sure there are no merge conflicts.

- Unit tests and integration tests are added to the code and all the tests are passing. Also, make sure that the code test coverage is as per the team standard.

- PR description is self-explanatory.

- Make sure the code follows the coding guideline standards.

Each team creates an extensive checklist that they can use within the team.

Good code reviewer

Once the PR has ticked all the checkboxes, now the task of code reviewer begins. The most important thing for a code reviewer is to make sure you are empathetic toward the person asking for a review. Here are a few good qualities/pointers I have generally seen in good code reviewers:

- **Be respectful**: The most important quality of a code reviewer is to be respectful while doing the code review. None of the comments should be personal or demeaning.

- **Be verbose**: Code reviewers who are verbose and communicate the reasoning behind their comments are always preferred by coders.

- **Dedicated time**: As the project grows, it is natural to have more developers join the team and more code review requests to come. Allocating time daily for code review helps keep the process streamlined and makes sure you are accounting for the time for code review while planning any work.

- **Talk in person**: If the code changes are not clear or the requested changes are large enough, it is better to talk to the person and discuss the changes rather than write comments and do too much back and forth.

- **Focus on core issues**: Code reviewers should primarily focus on core problems, corner cases, bugs, architectural issues, code readability, and maintainability. Other issues such as a bad variable name and typos are also important for the code reviewer to point out, but these should not be the primary code review agenda.

- **Reviewing test code**: While a lot of reviewers don't review test code, it is important to make sure the code reviewer focuses on reviewing the test code because test cases primarily help in catching bugs early on, and having a test coverage always improves the velocity of the development of any feature.

- **Security and vulnerability**: The code reviewer should always be asking and confirming whether any security and vulnerability changes have been introduced into the code and how they are being handled.

- **Have a code review checklist**: Having a code reviewer checklist is important so that the code reviewer is covering all the important pointers while reviewing the code.

Code review takes time, and it is a culture that needs to be built within the team. As a project and team matures, the code review process improves. It is important to make sure that the early members of the team are setting the right culture and code review process so that it is easy to follow as new members join the team.

Summary

In this chapter, we have learned how to use Git efficiently. We have explored different recommended good practices that can ease the SDLC and also help in writing better maintainable code. Git hooks can help us automate a lot of linting and we have learned how we can use Git hooks in our project. We also saw how we can use GitHub and GitHub Actions in the Django project, which can help us automate multiple workflows. Code review is a process that needs guidelines, and we have learned how to create different guidelines for code review. We have also seen different aspects of PR review and how to implement guidelines for PR review in the team to improve a good coding culture.

In the next chapter, we will explore how to deploy our Django code to the cloud using **Amazon Web Services** (**AWS**). AWS is the leading cloud service provider currently available in the market. We shall discuss more about AWS in our next chapter and learn how to deploy our Django application using AWS.

Part 4 –
Deploying and Monitoring Django Applications in Production

AWS is the leading cloud provider and is widely used in the industry to host web applications. In this part, we will learn how to deploy our Django application using AWS. We will use Amazon RDS, Amazon ElastiCache, AWS Elastic Beanstalk, and other AWS services to host our Django application in production. In this part, we will also learn how we can monitor our Django application. We will learn how to integrate different tools into our Django application, such as error monitoring tools (e.g., Rollbar) and **Application Performance Monitoring** (**APM**) tools (e.g., New Relic).

This part has the following chapters:

- *Chapter 13, Deploying Django in AWS*
- *Chapter 14, Monitoring Django Applications*

13

Deploying Django in AWS

In *Chapter 12*, we learned how to work with version control and use CI pipelines to run test cases and verify our code changes efficiently. In this chapter, we will take the next step toward our development cycle: learning how to deploy our Django application in production. While most books only cover application code development, we are going to take a step further and make sure we learn how to deploy our code on production. We are going to use **Amazon Web Services (AWS)** as a platform to deploy our Django code. There are several other cloud providers available, such as Azure, **Google Cloud Platform (GCP)**, and Digital Ocean, but out of all the major cloud providers, AWS has one-third of the market share. This is why we will learn how to use AWS to host our Django application.

> **Important note**
>
> AWS is a vast topic and is a book in itself. We shall not look at how AWS works in depth; rather, our primary focus will be to learn how we can deploy our Django application to AWS. If you or your company do not use AWS, then you can skip this chapter since this chapter is heavily coupled with AWS.

We will cover the following topics in this chapter:

- Learning the basics of AWS
- Integrating with AWS to deploy Django
- Deploying a Django application with GitHub Actions and Beanstalk
- Following the best practices while using AWS

Technical requirements

In this chapter, we shall work with AWS, primarily with Beanstalk, **Relational Database Service** (**RDS**), and ElastiCache, so we expect you to have some background in AWS. AWS provides free credits for 1 year that would be sufficient to follow all the steps mentioned in this chapter. You should also have access to a credit card so that you can create an account in AWS as you will need to enter your credit card details.

Please note that the examples used in this chapter fall within the free tier, so you will not be charged if you follow the instructions mentioned in this book.

If you are not familiar with AWS, it is recommended that you read the book *AWS for System Administrators*, by *Prashant Lakhera*, *Packt Publishing*, before proceeding with this chapter since we expect you to have some experience with cloud services.

There is another book called *AWS Certified Developer – Associate Guide*, by *Vipul Tankariya and Bhavin Parman*, *Packt Publishing*, that covers more extensive topics on how to get started with AWS for developers.

The code for this chapter can be found in this book's GitHub repository at `https://github.com/PacktPublishing/Django-in-Production/tree/main/Chapter13`.

Since this chapter deals with AWS integration, you might face some challenges, so please feel free to ask any questions in the Discord channel. The link to join the Discord channel can be found in this book's GitHub repository.

> **Important note**
> In this chapter, we shall use AWS; all the examples and screenshots were valid when we published this book. However, AWS keeps on making improvements and releasing new features that might make the instructions/screenshots obsolete, so I will try to update the README section of the GitHub code with the latest instructions. Please follow the instructions mentioned in this book's GitHub repository or ask in the Discord channel if you have any doubts.

Learning the basics of AWS

AWS is a cloud service that lets us run our web applications on the cloud. AWS has numerous services that can be used to run our applications on the web and scale them as traffic grows. In this section, we will cover the basics of all the AWS services that we will be using to deploy our Django application to production.

Creating an account in AWS

First, we must create a new account with AWS. Here are the simple steps we can follow to create an account with AWS:

1. Go to https://aws.amazon.com and go to the **Sign up** page.

2. Fill out the form, as shown in *Figure 13.1*:

Sign up for AWS

Explore Free Tier products with a new AWS account.

To learn more, visit aws.amazon.com/free.

Root user email address
Used for account recovery and some administrative functions

> hello@thetldr.tech

AWS account name
Choose a name for your account. You can change this name in your account settings after you sign up.

> thetldrtech

Security check

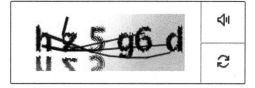

Type the characters as shown above

> hz5g6d

Verify email address

Figure 13.1: AWS sign up page

3. Go to your mailbox to get the six-digit OTP from AWS to enter on the next screen.

4. Fill in all the fields of the signup form regarding contact information and business information, as shown in *Figure 13.2*:

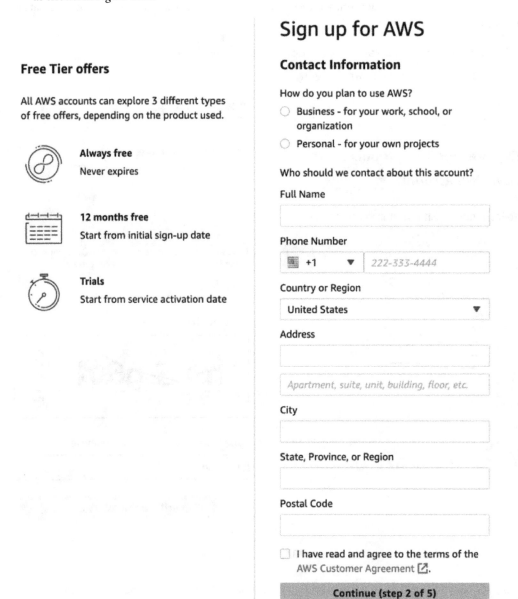

Free Tier offers

All AWS accounts can explore 3 different types of free offers, depending on the product used.

Always free
Never expires

12 months free
Start from initial sign-up date

Trials
Start from service activation date

Sign up for AWS

Contact Information

How do you plan to use AWS?

○ Business - for your work, school, or organization

○ Personal - for your own projects

Who should we contact about this account?

Full Name

Phone Number

⊞ +1 ▼ | 222-333-4444

Country or Region

United States   ▼

Address

| Apartment, suite, unit, building, floor, etc.

City

State, Province, or Region

Postal Code

☐ I have read and agree to the terms of the AWS Customer Agreement 🔗.

Continue (step 2 of 5)

Figure 13.2: AWS sign up contact information

5. Next, we have the **Billing information** page, as shown in *Figure 13.3*:

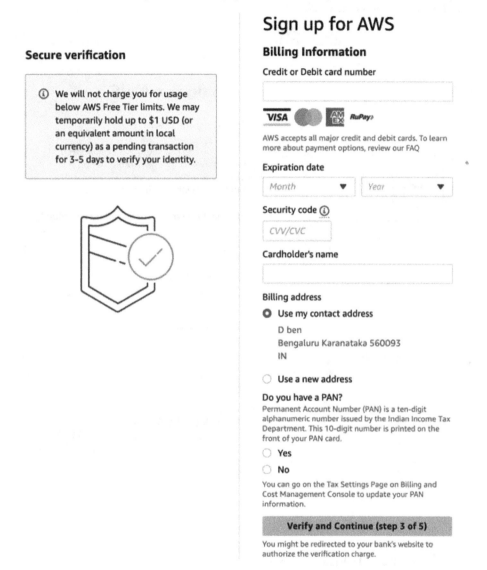

Figure 13.3: Billing information with credit card details for AWS sign up

Here, we have to enter our credit card details. AWS won't charge anything apart from any service we use. Since we will only be using the free tier service, we should not be billed any amount while following the instructions mentioned in this chapter. However, if you are concerned about the billing amount, then please create a *virtual credit card* by using services such as https://privacy.com and enter the credit card details from them; otherwise, use a debit card that has a smaller balance.

6.   Confirm your identity by providing a valid phone number; AWS will send a verification code to it. *Figure 13.4* shows the verification step:

Sign up for AWS

Confirm your identity

Before you can use your AWS account, you must verify your phone number. When you continue, the AWS automated system will contact you with a verification code.

How should we send you the verification code?

◉ Text message (SMS)

○ Voice call

Country or region code

India (+91)   ▼

Mobile phone number

Security check

Type the characters as shown above

Send SMS (step 4 of 5)

Figure 13.4: Entering a phone number to verify your identity

7.   Once your identity has been confirmed, you must select a support plan. For our learning purposes, we will select the **Basic support – Free** plan. Support plans are helpful when developers need help while working with AWS services. For any production application running in AWS, it is recommended to take business support or enterprise support to get faster solutions for any business-critical issues. In *Figure 13.5*, we have selected the **Basic support – Free** plan. This plan can be upgraded in the future:

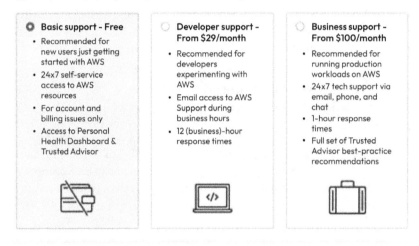

Sign up for AWS

Select a support plan

Choose a support plan for your business or personal account. Compare plans and pricing examples. You can change your plan anytime in the AWS Management Console.

Basic support - Free
- Recommended for new users just getting started with AWS
- 24x7 self-service access to AWS resources
- For account and billing issues only
- Access to Personal Health Dashboard & Trusted Advisor

Developer support - From $29/month
- Recommended for developers experimenting with AWS
- Email access to AWS Support during business hours
- 12 (business)-hour response times

Business support - From $100/month
- Recommended for running production workloads on AWS
- 24x7 tech support via email, phone, and chat
- 1-hour response times
- Full set of Trusted Advisor best-practice recommendations

Need Enterprise level support?
From $15,000 a month you will receive 15-minute response times and concierge-style experience with an assigned Technical Account Manager. Learn more.

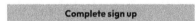

Complete sign up

Figure 13.5: Selecting an AWS support plan

With that, we have finished signing and have access to the AWS console. In the next section, we will gain a high-level overview of all the AWS services that we will use to deploy our application. We won't deep dive into any of the services but rather give a short description of what the AWS service is about and why we shall use this in our application. We'll start by learning about **Identity and Access Management (IAM)** in AWS.

Identity and Access Management

AWS provides an IAM service to manage all the authentication and authorization for the use of AWS services. Whenever we want to give access to any AWS service to any engineer or even any other application, we need to create new users in IAM. Let's look at the terminology that's used in IAM:

- **User**: A user in IAM can be a person or application that will be provided access to AWS resources. We should create a new user in IAM for every person and not share credentials. Creating IAM users is free.

- **Group**: A group in IAM is a collection of similar users. For example, let's say we want three users from customer support to have read-only access to AWS S3 and 10 users from the development team to have read/write access to AWS services. We can create two separate groups and add all the members to the respective groups. Now, every member in a group will have uniform access to AWS services.

- **Role**: A role in IAM is a combination of multiple IAM policies. A user can have multiple roles attached to their IAM user. For example, *EC2AdminRole* can have read and write access to the **Elastic Compute Cloud** (**EC2**) service, *RDSAdminRole* can have read and write access to RDS, and any IAM user with these roles attached would have both read and write access to both EC2 and RDS.

- **Policy**: A policy is a JSON document in AWS that lets us specify who has access to what resource and what kind of operation they are allowed to perform. Policy in AWS is the granular level of access control that is used alongside role, user, and user group to create a sophisticated access management system.

One more important aspect of IAM is setting up the AWS CLI and AWS SDK boto3 for any operation from the terminal. We can create a new **Access key** for a given user by going to the **Security Credentials** tab and clicking the **Create access key** button. *Figure 13.6* shows an example of an access key being created for a given user:

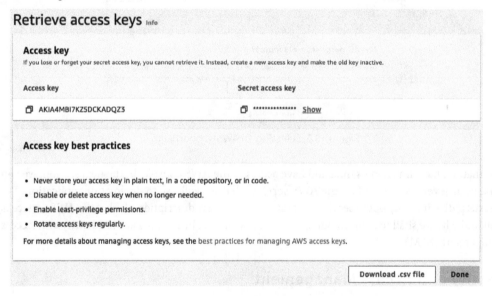

Figure 13.6: Using the AWS console to get an access key from IAM

Always keep access keys confidential. If an access key is ever exposed, please deactivate the key, and regenerate a new key as soon as possible. Hackers having an AWS access key means that they can control any AWS service in your account. AWS access keys should never be committed to the Git repository.

Next, let's learn about the most popularly used AWS service: EC2.

EC2

Amazon EC2 is a web service that provides secure, resizable computing capacity in the cloud. EC2 can be considered as the bare-metal server that can be used for any computational purpose. These are dedicated servers that can be provisioned as per your requirements. This is helpful as companies do not need to worry about the hardware if their product sees huge growth – they can focus on code application logic and let AWS take care of how the new servers can be provisioned. AWS has a standard and transparent pricing model for EC2 instances on per-second billing. This means you will only be charged for the time your EC2 instances are running. If they are deleted or stopped, then there is no additional cost.

EC2 instances can be considered the smallest building blocks of any web application. As our application grows, we can add more EC2 instances to scale our application. We use load balancers to put all such EC2 instances in a bucket.

Elastic Load Balancer (ELB)

ELB is a load balancing service by AWS that can distribute incoming traffic across multiple EC2 instances. Using a load balancer in our application can provide the application with high availability, better fault tolerance, and better scalability. AWS ELB primarily has three types of load balancers:

- **Classic load balancer**: This distributes incoming application traffic across multiple Amazon EC2 instances
- **Application load balancer**: This routes HTTP/HTTPS traffic based on content
- **Network load balancer**: This directs traffic at the transport layer, which is ideal for high-performance scenarios

We'll be using AWS Elastic Beanstalk to deploy our Django application, and Elastic Beanstalk recommends using an application load balancer. In the next section, we'll learn about AWS Elastic Beanstalk and why we should use it to deploy our Django application.

Elastic Beanstalk

Traditionally, whenever someone has to deploy a web application to the web, they have to configure different servers and services. A developer has to provision EC2 instances with the proper base image, configure **auto-scaling groups**, configure **security groups**, add a **load balancer**, and then deploy their application to the infrastructure. When we have multiple servers, we have to create a blue-green deployment or use services such as Code Deploy by AWS to make sure our code is deployed properly to the servers.

As per AWS's official documentation, Beanstalk is defined as follows:

> *"Elastic Beanstalk is a service for deploying and scaling web applications and services. Upload your code and Elastic Beanstalk automatically handles the deployment – from capacity provisioning, load balancing, and auto-scaling to application health monitoring."*

In this chapter, we'll be using Elastic Beanstalk to deploy our application to production.

Other services, such as AWS Lightsail, can be used for lighter scales. We have ECS and EKS, which are more sophisticated systems than Elastic Beanstalk with more control over the configurations. Elastic Beanstalk has a sweet spot that can handle considerable scale, and also not have too much configuration that it becomes overwhelming to the developers during initial setup. We'll discuss how to set up and configure Elastic Beanstalk with Django in the *Integrating AWS Beanstalk to deploy Django* section.

Now, let's learn how we can use Postgres in AWS.

RDS for Postgres

AWS RDS is a managed database service by AWS that can run Postgres, MySQL, MariaDB, Oracle, and SQL servers without the additional setups involved in setting up, operating, and scaling up the database. With a few additional clicks on the AWS console, developers can set up replication, have backups, upgrade database versions, and scale the server's size.

RDS takes away all the basic maintenance and infrastructure tasks of running a database server from the developers so that they can focus on application building. RDS also provides security features out of the box, such as how to connect to a database, encryption, and network isolation. AWS RDS also provides enhanced monitoring metrics that can help us analyze the performance of RDS servers.

AWS RDS has a free tier for 12 months, where we can use the RDS service free of cost – we'll be using the free tier service to follow along with the integration in this chapter (in the *Creating an AWS RDS Postgres instance and connecting to Django* section).

ElastiCache for Redis

AWS ElastiCache is a managed service by AWS that provides caching via Redis and MemCache. Similar to RDS, ElastiCache provides a scalable caching service where developers can set up a Redis server within a few minutes without getting too much into the hassle of server configuration. By using ElastiCache, we can easily upgrade our server, create replication instances, and perform maintenance using the AWS console. AWS ElastiCache has a free tier for 12 months. We'll use the free tier service in this chapter so that we can integrate AWS ElastiCache and Redis into our Django project.

Security groups and network components

Security groups are one of the most important concepts of AWS. Every service in AWS is isolated. To connect to any AWS service, such as RDS, ElastiCache, EC2 instances, and so on, we must configure security groups. In layman's terms, a security group is a firewall that can control the traffic to and from resources. We can create as many security groups as we want, and they are free of charge. We shall use the default security groups created by Elastic Beanstalk while integrating Django with Elastic Beanstalk in the *Integrating AWS Elastic Beanstalk to deploy Django* section.

AWS Secrets Manager

AWS Secrets Manager is a secrets management service that helps us process access to our applications, services, and resources. This helps us to easily rotate, manage, and retrieve database credentials, API keys, and other secrets that are vulnerable and should not be shared. Secrets Manager is a paid service with $0.40 per secret per month. AWS also provides a free 1-month trial. We'll utilize the 1-month free trial in this chapter to learn and explore how to integrate AWS Secrets Manager into our Django project. We will use Secrets Manager to save our application secrets and then retrieve them in the *Deploying a Django application using GitHub Actions in Elastic Beanstalk* section.

Route 53

AWS Route 53 is a **Domain Name System** (**DNS**) web service that can be used to purchase and maintain domains. Every consumer-facing application needs to have a domain name associated with it, and we can purchase domains from AWS Route 53 or transfer domains to Route 53 if the domain is purchased via another platform, such as Godaddy, Namecheap, and so on.

AWS Route 53 has a lot of added advantages. For example, it can be used for traffic load balancing, DNS failover, and other features that can help in integrating with other AWS services.

The AWS Billing console

The AWS Billing console allows us to easily understand the cost associated with all the AWS services. Developers need to learn how to use the AWS Billing console to monitor the costs associated with AWS services. There are 100+ services by AWS and it is common to have unused AWS resources running in an AWS account. AWS will only charge us for the resources that are running in the AWS account, so we can use the AWS Billing console to identify all the resources that are running.

CloudWatch

AWS CloudWatch is a monitoring service that can be used to monitor AWS services and applications running on AWS. AWS CloudWatch provides us with the ability to add alerts to our AWS resources. Most of the AWS resources have Cloudwatch metrics and are automatically integrated. All the logs that are pushed to CloudWatch from the AWS service can be queried using a CloudWatch log insights query.

With that, we've looked at the different services that are offered by AWS. It takes a significant amount of effort to orchestrate different AWS services and set up a production infrastructure to run Django. On the other hand, AWS Elastic Beanstalk can orchestrate the required AWS services to create a production environment within a few minutes. Let's learn how to use AWS Elastic Beanstalk to deploy a Django application.

Integrating AWS Elastic Beanstalk to deploy Django

In this section, we'll learn how to use AWS Elastic Beanstalk to deploy Django applications. Let's look at a basic Django example to help exemplify this.

Integrating Beanstalk with a basic Django app

First, follow the guide mentioned in *Chapter 1* to create a Django application, followed by the guide mentioned in *Chapter 11* to dockerize the Django application. Also, follow the instructions mentioned in `https://github.com/PacktPublishing/Django-in-Production/tree/main/Chapter13#integrating-beanstalk-with-basic-django-app`.

Before starting the Elastic Beanstalk integration, please ensure that the Django project is working by running `docker compose up`. We can access the Django admin page by going to `http://127.0.0.1/admin`:

```
docker compose up --build
```

Now that our local setup is running the Django application using Docker, let's focus on creating a new Beanstalk environment:

1. Install the Elastic Beanstalk (`eb`) CLI tool by using the following command:

    ```
    pip install awsebcli
    ```

    We'll be using the `eb` command to interact with and set up our application on Elastic Beanstalk.

2. After installing the CLI, we are ready to configure our project directory by running the `eb init` command. We need to select the AWS region where our application will be operating. Select the number corresponding to the region. Here, we will choose `Asia Pacific (Mumbai) - 6`:

    ```
    > eb init
    Select a default region
    1) us-east-1 : US East (N. Virginia)
    2) us-west-1 : US West (N. California)
    3) us-west-2 : US West (Oregon)
    4) eu-west-1 : Europe (Ireland)
    ```

```
5) eu-central-1 : Europe (Frankfurt)
6) ap-south-1 : Asia Pacific (Mumbai)
7) ap-southeast-1 : Asia Pacific (Singapore)
...
(default is 3): 6
```

3. Next, provide the access key and secret key so that the CLI can access the AWS account and manage/create resources. For more details on how to get AWS security credentials, please visit the official documentation by AWS at https://aws.amazon.com/blogs/security/wheres-my-secret-access-key/:

```
You have not yet set up your credentials or your credentials are
incorrect.
You must provide your credentials.
(aws-access-id): AKIAJOUAASEXAMPLE
(aws-secret-key): 5ZRIrtTM4ciIAvd4EXAMPLEDtm+PiPSzpoK
```

Please remember to not share AWS credentials with anyone. If they fall into the wrong hands, they can take over your whole AWS infrastructure.

4. Next, we have to give our application a name:

```
Enter Application Name
(default is "eb"): django-demo-app
Application django-demo-app has been created.
```

5. Elastic Beanstalk can run on multiple platforms. Our application is a Django application running in Docker, so we have to select the Docker platform:

```
Select a platform.
1) .NET Core on Linux
2) .NET on Windows Server
3) Docker
4) Go
5) Java
6) Node.js
7) PHP
8) Packer
9) Python
10) Ruby
11) Tomcat
(make a selection): 3
```

6. Select the Docker platform you want Beanstalk to run on. We'll choose `Docker running on 64-bit Amazon Linus 2023`:

```
Select a platform branch.
1) Docker running on 64bit Amazon Linux 2023
2) Docker running on 64bit Amazon Linux 2
3) ECS running on 64bit Amazon Linux 2
(default is 1): 1
```

7. Next, Beanstalk automatically sets up SSH access for our AWS instances. We need to choose yes to assign the SSH key pair. This will allow us to directly SSH to our Beanstalk application:

```
Do you want to set up SSH for your instances?
(y/n): y
```

8. Beanstalk creates a new SSH key pair that it will use to give access to EC2 instances:

```
Type a keypair name.
(Default is aws-eb):
Generating public/private rsa key pair.
Enter passphrase (empty for no passphrase):
Enter same passphrase again:

Your identification has been saved in /Users/argo/.ssh/aws-eb
Your public key has been saved in /Users/argo/.ssh/aws-eb.pub
```

With that, our Beanstalk application has been created. Now, we can go to the AWS console and find our Beanstalk application ready. *Figure 13.7* shows our newly created Beanstalk application:

Figure 13.7: The AWS console showing the newly created Beanstalk application, django-demo-app

When we go inside the `django-demo-app` application in the AWS console, we won't see an environment, as depicted in *Figure 13.8*. This is because we have created the Beanstalk application but have not created any environment that will run the Django application on Beanstalk:

Figure 13.8: The AWS console showing the empty Beanstalk application

9. To create the environment, go inside the Django application folder and run the `eb create <environment-name>` command. For example, if we name our environment `prod-env`, then we should use the following command to create a Beanstalk environment:

```
eb create prod-env
```

Now, we can go to the AWS Elastic Beanstalk dashboard to see our Beanstalk environment being created. This step generally takes 10 minutes since Elastic Beanstalk will create all the required AWS components, such as an EC2 instance, **Application Load Balancer** (**ALB**), auto-scaling groups, CloudWatch alarms, and so on:

Figure 13.9: The AWS Elastic Beanstalk dashboard showing the prod-env environment

Once the Elastic Beanstalk environment is ready, we will see the **Health** status go from **Pending** to **Ok**. This indicates that the application is ready to serve traffic. In *Figure 13.10*, we can see the status of `prod-env`, which states that it is ready to serve traffic:

Figure 13.10: AWS Elastic Beanstalk prod-env showing a health status of Ok

The **Domain** value shown in *Figure 13.10* is the public domain that can be used to access our Django application from the internet. In our case, it is `http://prod-env.eba-e5mff922.ap-south-1.elasticbeanstalk.com/admin` (dummy URL); when we open it on our browser, we should be able to see the initial Django web page. We can see the browser's output in *Figure 13.11*:

Figure 13.11: Initial Django web page

The `eb` CLI will only work for projects that are in Git and have at least one commit. If you face an error, as shown here, please make sure your project is being tracked using Git:

```
> eb create prod-env
WARNING: Git is in a detached head state. Using branch "default"
WARNING: Git is in a detached head state. Using branch "default"
WARNING: Git is in a detached head state. Using branch "default"
WARNING: Git is in a detached head state. Using branch "default"
ERROR: CommandError - git could not find the HEAD; most likely because
there are no git commits present
```

Using our boilerplate Django project code, we have seen how to deploy our application to Elastic Beanstalk. In our example, we have used basic boilerplate code that deploys a Django application without connecting to the database or any other external service needed to run the web service. Now, let's learn how to connect the Django application to AWS RDS and ElastiCache.

Creating an AWS RDS Postgres instance and connecting to Django

Elastic Beanstalk provides us with the option to create RDS for a given Elastic Beanstalk environment. However, we won't use that option to ensure that our RDS instance is decoupled from Beanstalk.

Now, let's learn how to create an RDS instance that can be connected to our Beanstalk environment and configure our Django code.

Creating a new RDS environment

Here are the simple steps we can follow to create a new RDS environment:

1. Go to the RDS dashboard in the AWS console:

Figure 13.12: Navigating to RDS on the AWS dashboard

2. Go to the **Databases** section on the left sidebar, then click on **Create Database** to open the wizard, as shown in *Figure 13.13*:

Figure 13.13: The RDS dashboard in the AWS console

3. Select the **Easy Create** option and then select the following options (alternatively, you can configure the database with more options by using the **Standard Create** option. We'll follow the **Easy Create** option in our tutorial):

    A. Select **Option** in the wizard. Then, select **PostgreSQL** under **Engine Type**.
    B. Select **Free tier**.

C.   Name the database. We'll name it django-demo.

D.   Specify a master username. We'll use django_db_user.

E.   Specify the master password. We'll use abc^A12a*12.

F.   Then, click the **Create** button.

4.   It will take RDS 4-5 minutes to create the database. Once it has, we'll see that the status of the database is **Creating**:

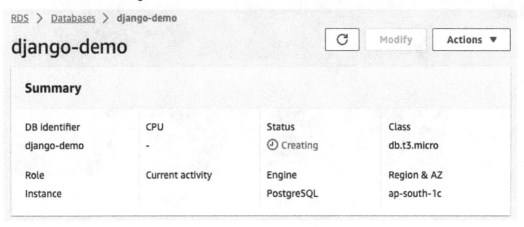

Figure 13.14: A new RDS instance being created

With that, we have created the database. Next, we need to connect our database to Django code. By default, RDS is not exposed to the internet, which means we cannot connect to the RDS instance outside of our AWS VPC network. It is a good practice to not expose the RDS server outside to the internet.

However, this doesn't mean our Elastic Beanstalk servers can directly connect to the RDS instance. To connect to the RDS instance, we have to configure our VPC security groups to enable connections.

Connecting RDS to the Django server

Here are the steps to connect RDS to the Django server:

1. Go to the **Elastic Beanstalk** dashboard and select the Django application we created earlier. Then, select the `prod-env` option and select the **Configuration** option from the left sidebar. This will show us all the configurations for our instance:

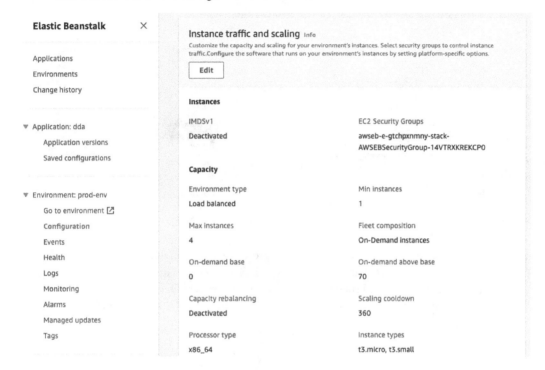

Figure 13.15: Elastic beanstalk configuration for prod-env we created earlier

2. Copy the EC2 security group shown – in our case, it is `awseb-e-gtchpxnmny-stack-AWSEBSecurityGroup-14VTRXKREKCP0`.

3. Now, go back to the RDS console and select the created RDS instance. In the **Connectivity & security** section, click on the **VPC security groups** value – in our case, it is `default (sg-5acc5623)`:

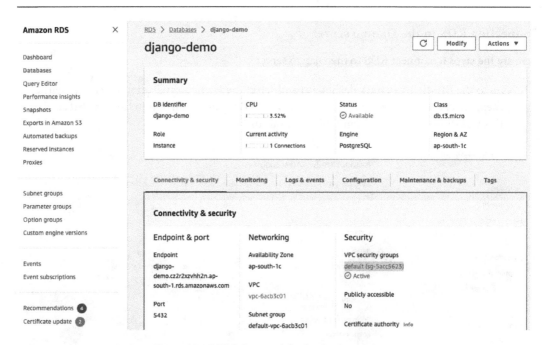

Figure 13.16: RDS Connectivity & security settings

4. The `default (sg-5acc5623)` security group controls which AWS resource will have access to the RDS instance. We need to allow our EC2 instance to connect to the RDS instance. We have to add new inbound rules to this security group. By default, we won't have any inbound rules; to add new inbound rules, we should click on the **Edit inbound rules** button at the bottom right, as shown in *Figure 13.17*:

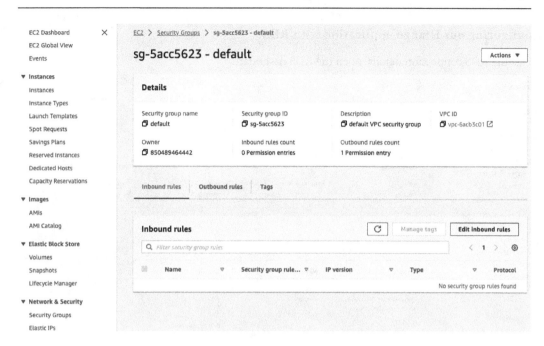

Figure 13.17: RDS default security group

5.   Set **Type** to **PostgreSQL** and **Source** to **Custom**, then search for the security group, copy and paste the same value we selected in *Step 2*, and select the security group:

Figure 13.18: Updating inbound rules to allow a connection from Elastic Beanstalk to RDS

This will enable our EC2 instances running inside Elastic Beanstalk to connect to the RDS instance. Now, let's configure our Django application code with the RDS connection details.

Configuring our Django application with RDS

To get the RDS connection details, open the RDS dashboard:

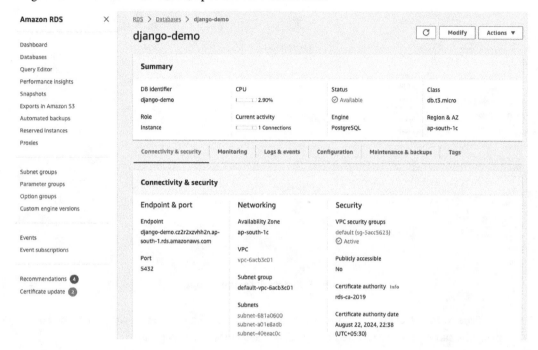

Figure 13.19: RDS dashboard showing the connection configuration

Find the **Endpoint & port** from the dashboard, as shown in *Figure 13.19*. Here's the configuration we need to integrate RDS with Elastic Beanstalk:

- **Endpoint**: django-demo.cz2r2xzvhh2n.ap-south-1.rds.amazonaws.com
- **Port**: 5432
- **Username**: django_db_user (from *Step 3D* of the *Creating a new RDS environment* section)
- **Password**: abc^A12a*12 (from *Step 3E* of the *Creating a new RDS environment* section)

Update the Django .env file with the connection string:

```
DEBUG=True
DJANGO_ALLOWED_HOSTS=*
DB_ENGINE=django.db.backends.postgresql
DB_NAME=postgres
DB_USERNAME=django_db_user
DB_PASSWORD=abc^A12a*12
DB_HOSTNAME=django-demo.cz2r2xzvhh2n.ap-south-1.rds.amazonaws.com
```

```
DB_PORT=5432
REDIS_HOST=redis_db
REDIS_PORT=6379
REDIS_PASSWORD=redisPassWord
```

We also have our Django-health-check package configured in settings.py to ensure our healthcheck endpoint can test and validate the database connection.

> **Important note**
>
> We have hardcoded the RDS connection value in our Django setting files for testing purposes. RDS connection values are sensitive information and should not be shared or saved in the application code. We'll learn how to pass such sensitive information in the *Deploying a Django application using GitHub Actions in Elastic Beanstalk* section.

Our code is ready to be deployed and tested. Now, we need to move our Django settings file to Git staged. This is essential since the eb CLI will only deploy code that is Git tracked or is in the staged state when we pass a special flag, --stage:

```
> git add .env
> eb deploy --staged
Creating application version archive "app-8e23-231024_182626309787-
stage-231024_182626309839".
Uploading dda/app-8e23-231024_182626309787-stage-231024_182626309839.
zip to S3. This may take a while.
Upload Complete.
2023-10-24 12:56:27    INFO    Environment update is starting.
2023-10-24 12:56:31    INFO    Deploying new version to instance(s).
2023-10-24 12:56:46    INFO    Instance deployment completed
successfully.
2023-10-24 12:56:52    INFO    New application version was deployed to
running EC2 instances.
2023-10-24 12:56:52    INFO    Environment update completed
successfully.
```

Our Django code can connect to the database. To verify this, open the same Elastic Beanstalk environment URL – http://prod-env.eba-e5mff922.ap-south-1.elasticbeanstalk.com/admin. You should be able to see the same page as what's shown in *Figure 13.11*.

If the page is showing 502, then Django can't connect to the RDS instance – there can be multiple reasons for this. Please check the **Troubleshooting** README section in this book's GitHub repository at https://github.com/PacktPublishing/Django-in-Production/tree/main/Chapter13#troubleshooting.

With that, we have explored the basics of how to create a new RDS and connect it to our Django service to our RDS instance. Now, let's learn how to connect our Django service to our Redis service.

Creating an ElastiCache Redis instance and connecting it to Django

ElastiCache provides us with a managed Redis cluster that we can use for caching and managing our task queue for the celery service. In this section, we'll create our Redis cluster. We'll use our basic configuration to create the Redis server. We will create a basic cluster for demo purposes. To do so, follow these steps:

1. Go to the ElastiCache dashboard and select **Create Redis Cluster**.

2. In the wizard, select the **Design your own Cluster** option.

3. Choose **Creating method – Cluster Cache** and select the **Cluster mode – Disabled** configuration.

4. Choose a name for the cluster; we'll name it django-demo-cache.

5. Uncheck **Multi-AZ and Auto-failover**.

6. In the cluster settings, make the following changes:

    A. Set **Engine version** to 7.0 or more

    B. Set **Port** to 6378

    G. Set **Node type** to cache.t4g.micro

    H. Set **Number of replicas** to 0

7. Keep **Connectivity** and **Availability Zone placements** as-is.

8. In the **Security** section, uncheck **Encryption in transit**.

9. Open a new tab and go to the AWS console's **Security groups** dashboard. Then, create a new security group that will be attached to the Redis instance.

10. Just like the RDS security group, we need to create a new inbound rule that will expose port 6379. Attach the Beanstalk security group as **Source – Custom** and select the security group from the dropdown, as shown in *Figure 13.20*:

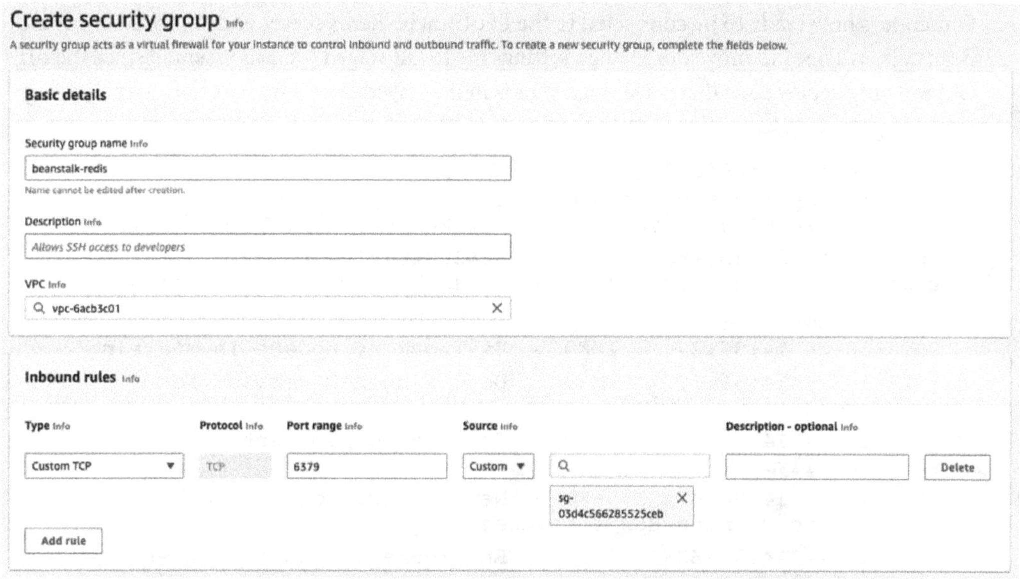

Create security group Info

A security group acts as a virtual firewall for your instance to control inbound and outbound traffic. To create a new security group, complete the fields below.

Basic details

Security group name Info

```
beanstalk-redis
```
Name cannot be edited after creation.

Description Info

```
Allows SSH access to developers
```

VPC Info

```
Q  vpc-6acb3c01                                    ×
```

Inbound rules Info

Type Info	Protocol Info	Port range Info	Source Info		Description - optional Info	
Custom TCP ▼	TCP	6379	Custom ▼	Q		Delete
				sg-03d4c566285525ceb ×		

Add rule

Figure 13.20: Creating a new security group

11. Now, go back to the Redis wizard. Then, in the **Selected Security Groups** section, select the newly created security group we created in *Step 10*.

12. Now, select the **Create** option. It will take 5-6 minutes to create a Redis instance that is ready for connection.

   Once the Redis cluster is **Available**, select **Primary endpoint**, and then configure the Django .env file with the Redis cache:

```
DEBUG=True
DJANGO_ALLOWED_HOSTS=*
DB_ENGINE=django.db.backends.postgresql
DB_NAME=postgres
DB_USERNAME=django_db_user
DB_PASSWORD=abc^A12a*12
DB_HOSTNAME=django-demo.cz2r2xzvhh2n.ap-south 1.rds.amazonaws.
com
DB_PORT=5432
REDIS_HOST=django-demo-
cache.dysbo0.ng.0001.aps1.cache.amazonaws.com
REDIS_PORT=6379
```

Our code is now ready to be connected to the ElastiCache Redis server and deployed to Elastic Beanstalk. We need to move our Django settings file to Git staged – this is essential since the eb CLI will only deploy code that is Git tracked or is in the staged state if we pass the --stage flag:

```
> git add .env
> eb deploy --staged
Creating application version archive "app-8e23-
231024_202626040031-stage-231024_202626040080".
Uploading dda/app-8e23-231024_202626040031-
stage-231024_202626040080.zip to S3. This may take a while.
Upload Complete.
2023-10-24 14:56:27    INFO    Environment update is starting.
2023-10-24 14:56:30    INFO    Deploying new version to
instance(s).
2023-10-24 14:56:47    INFO    Instance deployment completed
successfully.
2023-10-24 14:56:52    INFO    New application version was
deployed to running EC2 instances.
2023-10-24 14:56:52    INFO    Environment update completed
successfully.
```

13. Open the /ht/ endpoint in your browser to verify the database and cache connection. In *Figure 13.21*, we can see the expected output when our RDS Postgres and ElastiCache Redis instances are properly configured and connected to the Django server:

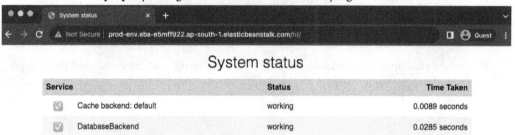

Figure 13.21: The health check endpoint, /ht/, showing success

With that, we have learned how to connect our RDS Postgres and ElastiCache Redis cluster to the Django service. We used a very simple setup to create our databases and connect to the Django service. We will discuss all the best practices on how to use RDS and Elasticache in production in the *Following the best practices for the AWS infrastructure* section.

We learned how to integrate Celery with Django in *Chapter 8*, while in *Chapter 11*, we saw how to use Docker to run Celery workers alongside Celery beat. We used Docker to run our Django application in the Elastic Beanstalk environment. For small-scale applications, we can use the same setup and run Celery.

For large-scale applications, we should decouple the web application running Django and workers running Celery tasks. We need to set up different Beanstalk environments that can run the Celery and Celery beat workers. Setting up a Beanstalk environment for workers is beyond the scope of this book.

Now, let's learn how to deploy our Django application with GitHub Actions.

Deploying a Django application using GitHub Actions in Elastic Beanstalk

In *Chapter 12*, we learned how to use GitHub Actions to run test cases. Now, we'll learn how to deploy our Django application in Elastic Beanstalk using GitHub Actions.

First, let's learn how to connect GitHub Actions to our AWS account.

AWS has official GitHub Actions properties (`https://github.com/aws-actions/configure-aws-credentials`) that take care of authentication using an AWS access key and AWS secret access key, both of which we can get from the AWS IAM console. We need to extract the secret key and add the secret key to where we want to run the actions in the respective GitHub repository.

> **Important note**
> While creating the AWS access key, it is advisable to create a separate IAM user with limited access to AWS resources so that it can be used by GitHub Actions.

Once we have the AWS access key, we can follow the GitHub Actions documentation. GitHub has a very good guide on how to use secrets in GitHub Actions. Please go through the guide at `https://docs.github.com/en/actions/security-guides/using-secrets-in-github-actions`.

Once we have our secrets configured in GitHub, we can test our GitHub-AWS setup using the following example. Here, we are using the AWS CLI to verify whether GitHub Actions can access our AWS account:

```
name: Test Github-AWS actions integration
on: [workflow_dispatch]
jobs:
  build:
    runs-on: ubuntu-latest
    strategy:
      matrix:
        python-version: [3.8]
    steps:
      - uses: actions/checkout@v1
      - name: Configure AWS Credentials
        uses: aws-actions/configure-aws-credentials@v4
```

```
    with:
      aws-access-key-id: ${{ secrets.AWS_ACCESS_KEY_ID }}
      aws-secret-access-key: ${{ secrets.AWS_SECRET_KEY }}
      aws-region: ap-south-1
  - name: get caller identity
    run: aws sts get-caller-identity
```

With that, we have verified that GitHub Actions can connect to the AWS account. In the *Connecting RDS to the Django server* section, while integrating RDS with Django, we were hardcoding all the RDS secret values in our code. That is a bad practice and also a security hazard since our secrets would be available in the code. Therefore, we should pass such sensitive values using environment variables and populate the environment variable using AWS Secrets Manager. We can use the official GitHub Actions by AWS to extract our secrets and save them in a .env file. In the following example, we are downloading all the secrets that start with beta:

```
- name: Get Secret Names by Prefix
  uses: aws-actions/aws-secretsmanager-get-secrets@v1
  with:
    secret-ids: |
      django*    # Retrieves all secrets that start with 'django'
```

When we run GitHub Actions in the preceding code block – let's say we have three secrets that start with django, (djangoSecretName, djangoTest, and django/NewSecret) – then three environment variables would be created, as follows:

```
DJANGOSECRETNAME: secretValue1
DJANGOTEST: secretValue2
DJANGO_NEWSECRET: secretValue3
```

For more details about how to use AWS Secrets Manager GitHub Actions, check out the official documentation at https://github.com/aws-actions/aws-secretsmanager-get-secrets.

To save the environment variables to a particular file, do the following:

```
- name: Save secret manager variables to .env
  run: printenv | grep "^DJANGO" > .env
```

The preceding code will save all the environment variables starting with DJANGO in the .env file. We can access this .env file using the Docker environment.

With that, we have learned how to deploy new application code using the eb CLI from our local system. Next, we'll learn how to move the same operation to GitHub Actions.

Here is a small example of all the steps that are needed to deploy our application code to Elastic Beanstalk using GitHub Actions:

```
- name: Install aws cli and beanstalk cli
  run: pip install awsebcli awscli

- name: Get Secret Names by Prefix
  uses: aws-actions/aws-secretsmanager-get-secrets@v1
  with:
    secret-ids: |
      django*    # Retrieves all secrets that start with 'beta'
- name: Save secret manager variables to .env
  run: printenv | grep "^DJANGO" > .env

- name: git add for .env
  run: git add .

- name: Run eb init
  run: eb init django-demo-app -region ap-south-1 -platform docker

- name: Run eb use
  run: eb use prod-env -region ap-south-1

- name: Run eb deploy
  run: eb deploy -staged
```

Now, we can trigger the deployment of our Django code from GitHub Actions. We won't cover each step mentioned in the code in depth since we discussed them in the *Integrating AWS Elastic Beanstalk to deploy Django* section.

We can configure the Beanstalk environment further using the Elastic Beanstalk AWS console dashboard. For more details, please read the official documentation at `https://docs.aws.amazon.com/elasticbeanstalk/latest/dg/command-options.html`.

Elastic Beanstalk also allows developers to configure the environment programmatically using `.ebextensions`. Elastic Beanstalk is a powerful platform that gives us a lot of control over how to configure our application. The AWS documentation is the best resource to gather all the information about how to configure AWS Elastic Beanstalk and what to configure.

> **Further reading**
>
> We are currently running our Django application in development mode so that you can easily follow along with all the examples. As we want to scale our application to production, we need to add `nginx` and `gunicorn`, and set up the application, depending on the requirements. Every application has a scale and requirements. To learn more, please read the AWS documentation and set up your application as per your requirements and scale.

In the next section, we'll discuss the best practices to follow while working with AWS services.

Following the best practices for the AWS infrastructure

In this section, we'll learn about the best practices that should be followed while working with AWS services. AWS has a good architecture framework that AWS recommends organizations follow. For more details, check out `https://aws.amazon.com/architecture/well-architected/`.

The AWS Well-Architected Framework is extensive and might need additional effort to reach the state of. We won't dive too deep into the AWS Well-Architected Framework, instead just listing the best practices that can help developers configure their AWS services better and make sure there are no loopholes in the system. We'll start by looking into the best practices for RDS.

Best practices for RDS

In this section, we'll discuss all the best practices for RDS that we should follow to run RDS in production:

- **Enable Cluster Delete protection**: Enabling this flag in production is a must so that we do not accidentally delete RDS clusters.

- **Enable Multi-AZ**: The database is the most important component in an application as it saves all the persistent information that's important for the business. By enabling RDS Multi-AZ, we ensure our RDS server has high availability/failover. If any of the AWS zones become unavailable, then RDS will automatically switch over to another zone.

- **Performance Insights**: Using Performance Insights, we can fine-tune the database's performance tuning and monitoring features. This helps us to quickly assess the load on the RDS server and also helps non-experts detect performance problems.

- **Enable RDS automated backups during low I/O time**: Enabling automatic backup for RDS is important. We should schedule the RDS backup during low usage time for the app – for example, post-midnight for consumer-facing apps.

- **Do not use the default port**: Do not use the default port of 5432 so that attackers and hackers can't identify the actual port.

- **RDS is not publicly accessible**: Ensure the RDS server is not accessible publicly. It should only be accessible within the AWS network, not directly through the internet.

- **Monitoring with CloudWatch**: AWS RDS has a lot of default CloudWatch metrics that are pushed to AWS CloudWatch. We can use CloudWatch to monitor RDS better and set alarms on appropriate thresholds.

- **Encrypt RDS data at rest**: Always ensure that RDS encrypts the database with AWS **Key Management Service (KMS)**.

Best practices for ElastiCache

In this section, we'll cover the best practices we can follow while working with ElastiCache:

- **Enable Multi-AZ clusters**: Enabling ElastiCache Multi-AZ clusters will help us have a higher degree of availability. AWS will guarantee a higher **service-level agreement (SLA)**. Whenever our primary node goes down, AWS will automatically perform a failover.

- **Do not use the default port**: Do not use the default port of 6379 for ElastiCache.

- **Monitoring with CloudWatch**: Setting up CloudWatch monitoring for the Redis cluster would help us in setting up alerts for our ElastiCache cluster. By default, ElastiCache provides host-level metrics (for example, CPU usage) and cache engine-specific metrics (for example, cache hits, misses, and so on).

Best practices for Elastic Beanstalk

Now, let's look at the best practices for Elastic Beanstalk:

- **Protect the network**: EC2 instances running Elastic Beanstalk should not be accessible outside via the internet. Adding ALB and making sure all the traffic is routed through the ALB would help keep the ALB in the public subnet while the EC2 instances run in a private subnet.

- **Least privilege access to IAM**: Elastic Beanstalk will use a service account role to perform different operations. We should give limited privileges to the service account role.

- **Rolling with an additional batch**: AWS's Beanstalk deployment policy can help in controlling how our application is deployed. Rolling with an additional batch is the safest option since we do not remove the existing fleet of machines and add new servers with the latest application code.

- **Set up CloudWatch metrics and monitoring**: Set up alarms for key Elastic Beanstalk metrics.

Best practices for IAM and security

Here are the best practices for IAM and security:

- **Never use the root account**: Never use the root account for any operation. The root account should never be used unless there is an absolute requirement that cannot be done by other IAM users.

- **Mandate two-factor authentication (2FA)**: Mandating 2FA for all AWS IAM human users would ensure additional security.

- **Apply for least-privilege permissions**: Whenever you're creating IAM policies, always be explicit regarding what permission to provide rather than giving blanket permission.

- **Regularly review and remove unused users and roles**: Conduct continuous monitoring and remove redundant users and roles.

With that, we have learned about the best practices we should follow while working with AWS components. We have tried to give a comprehensive view and list all the best practices in one place. However, it is strongly recommended that you follow the official AWS documentation before moving your application to production.

Summary

In this chapter, we learned about different AWS components that we need to use to deploy our Django application to production. We used AWS Elastic Beanstalk to configure and deploy our Django application in production. We also used AWS RDS and ElastiCache for database and cache purposes in our Django application. AWS ensures we can scale our application infrastructure easily.

AWS Elastic Beanstalk is a managed platform service that can orchestrate applications by helping us deploy and scale them. AWS encapsulates all the important configurations and links between AWS services that are needed to deploy and run our application at scale.

Running a stable and performant application is important. In the next chapter, we'll learn how we can continuously monitor our Django application running in production using tools such as New Relic, Rollbar, and others.

Monitoring Django Application

In *Chapter 13*, we learned how to deploy a Django application to AWS and run code in production. Now comes the last and most important part of application development – monitoring and maintenance.

Developing an application is the first step of product development. Users will use a product only when there is enough trust in the product. The first step toward establishing trust in a product is making the product stable. To achieve stability in our application, we need to have a good monitoring system and reduce errors and downtime. We shall learn how to use different tools to monitor our Django application.

In this chapter, we shall cover the following topics:

- Integrating error monitoring tools into a Django application
- Integrating uptime monitoring tools into a Django application
- Integrating APM tools into a Django application
- Integrating messaging tools into the development process
- Handling production incidents better
- Using blameless RCA for incidents

Technical requirements

In this chapter, we shall focus on integrating different third-party tools into our Django application. We expect you to be well-versed in the concepts discussed in the previous chapters. You should have basic knowledge of exception/error monitoring and **Application Performance Monitoring** (APM) tools and how they can be used to improve the stability and performance of applications. You are also expected to know about the basic concepts of uptime monitoring and should be using messaging tools such as Slack/Teams.

Here is the GitHub repository that has all the code and instructions for this chapter: https://github.com/PacktPublishing/Django-in-Production/tree/main/Chapter14

Integrating error monitoring tools

When we develop an application, developers will try to handle all the corner cases and write as much error-free code as possible. But, somehow, a few corner cases will be missed. A few errors will always slip by, and users will see occasional errors while using the service. These application errors are occasional, but it's important to address them to have a stable application. While working on a local development setup, these errors can be easily detected in the terminal. But when we move to production, it becomes difficult to detect these errors. A lot of beginners still use logs to detect raised exceptions. Error monitoring tools are lifelines to detect any production exceptions raised.

Tools such as Sentry, Rollbar, BugSnag, and so on are error/exception monitoring tools that help us track and fix exceptions raised in production. In this chapter, we shall use Rollbar (`https://rollbar.com/`) and integrate it into our Django project.

Integrating Rollbar into a Django project

To integrate Rollbar into our Django project, we will follow these steps:

1. Create an account in Rollbar (`https://rollbar.com/`) and use the wizard to create a new project. In *Figure 14.1*, we are creating a new project by giving the project name `DjangoDemoApp`:

Let's Create a New Project

Data in your Rollbar account is organized into Projects. Each project represents a single deployable/release-able service or application. Learn more

Create new project		
+ Select SDK	**PROJECT NAME** ❓	DjangoDemoApp
+ Integrate SDK		
	TIMEZONE ❓	Asia - Calcutta UTC+5.5
	TEAM ❓	Everyone 3 members

Cancel   Create Project   ⓘ Support

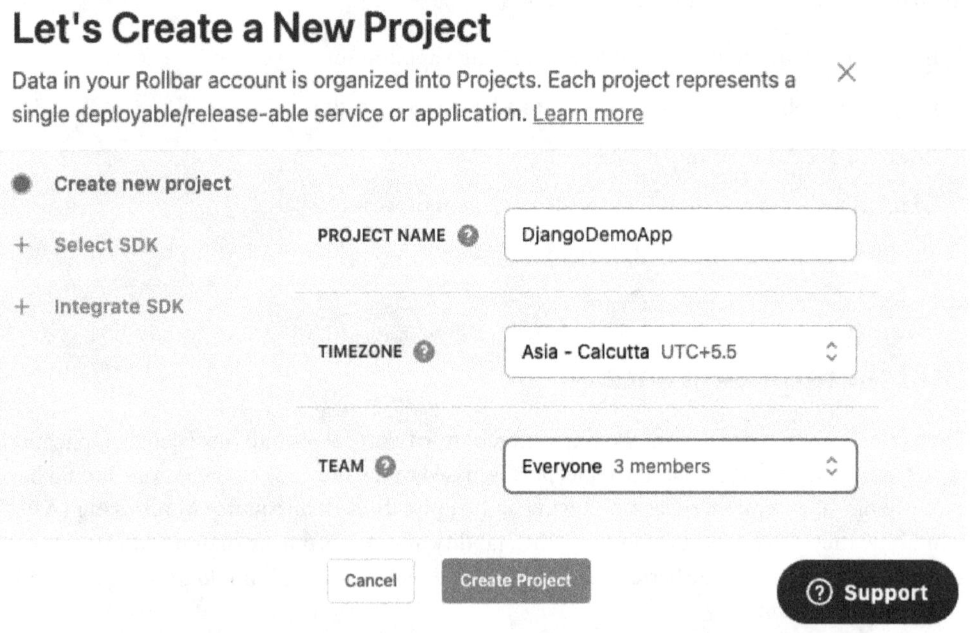

Figure 14.1: Wizard to create a new Django integration for Rollbar

2. Once we click the **Create Project** button, we will then move on to the next step, where we need to choose the framework. In *Figure 14.2*, we are selecting the **Django** framework:

Figure 14.2: Select the Django framework in the wizard

3. Once we click the **Continue** button, Rollbar gives us a set of instructions to follow, as mentioned in the next steps.

4. Install the `rollbar` SDK:

```
pip install rollbar
```

5. Configure the `rollbar` SDK for the Django project. Go to the `settings.py` file and add the Rollbar middleware to `MIDDLEWARE`:

```
MIDDLEWARE = [
    # OTHER MIDDLEWARES 'rollbar.contrib.django.middleware.
RollbarNotifierMiddleware',
]
```

6. Add the Rollbar access token in the `settings.py` file:

```
ROLLBAR = {
    'access_token': '<token>',
    'environment': 'dev' if DEBUG else 'prod',
    'root': BASE_DIR
}
```

We have integrated Rollbar into the Django application. To test the Django Rollbar integration, we need to intentionally create an exception error in our code. Add the following code in the Django `urls.py` file:

```python
from django.http import HttpResponse

def rollbar_test_view(request):
    a = None
    a.hello() # This would raise an exception
    return HttpResponse("Hello World!")

urlpatterns = [
    # old url config
    path('rollbar-test/', rollbar_test_view),
]
```

Now, when we open the URL in our browser, `http://127.0.0.1:8000/rollbar-test/` is mapped to the `index` view where we have introduced the error. We will see an exception in our browser and also see our first exception captured on the Rollbar dashboard. *Figure 14.3* shows the error capture in the Rollbar dashboard:

Figure 14.3: First error captured in Rollbar

Next, we will integrate Rollbar with Slack.

Integrating Rollbar with Slack

Rollbar has official documentation that mentions all the steps needed to integrate it into Slack. Follow the official guide at `https://docs.rollbar.com/docs/slack`.

Important note

We are not adding the integration steps in the current chapter because third-party tools are generally updated frequently.

Once Slack is integrated into Rollbar, whenever a new error is captured, we will see the error on the configured Slack channel. *Figure 14.4* shows an example error message posted by Rollbar:

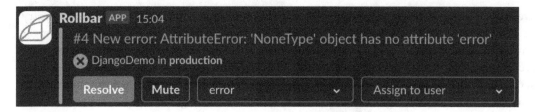

Figure 14.4: Slack notification for error captured in Rollbar

Now, we have configured Rollbar to capture all the errors happening in production and get notified on our Slack channel whenever a new exception is raised in production. The advantage of using Rollbar and similar tools is that it can capture the input that caused the exception to happen. This way, developers have more data on what caused errors and can fix issues faster rather than doing guesswork about the reason for the exception in production.

Best practices while working with error monitoring tools

As the application grows, the number of errors also might increase. Teams must follow best practices with error monitoring tools so that they don't get overwhelmed. Let us look into a few good practices while working with error monitoring tools:

- **Never silence errors forever**: It might be possible that the team is so occupied with new feature development that a particular customer-facing exception is taking a backseat. The team should always come back to an error and fix it when the workload has reduced rather than marking the issue as muted, which will make the error silent forever unless someone visits it.

- **Dedicate time to triage the dashboard**: Have a dedicated person in the team who will spend time triaging and assigning issues from Rollbar. This role can be on a rotation basis every month.

- **Connect source code**: Connecting source code to error monitoring systems can help in getting better stack trace and deployment information.

- **Use proper notification rules**: Notification rules are important for any monitoring tool. We don't want to lose out on important alerts, nor do we want to make too much noise. Each team/company should discuss internally creating a standard notification cadence.

Using Sentry

Sentry (`https://sentry.io/`) is an open source alternative error-monitoring tool that is leading the industry. Sentry has many features that provide automation, such as auto assign issues rule, which means when an error happens for a particular segment of the code, a developer would be automatically assigned the sentry issue without any manual intervention. We have used Rollbar primarily due to its generous free tier and free Slack integrations, which are paid on most of the other platforms.

I would prefer to use Sentry if the team size is bigger, and you are willing to pay for Sentry. Integrating Sentry is a straightforward step, the same as what we have done for Rollbar integration.

We have learned how to monitor errors that occur in production. Tools such as Rollbar and Sentry help us to track all code-level errors due to unexpected input or corner cases that developers might have missed. Such issues cause 5xx errors only for a particular number of requests. But there are scenarios where the whole system can go down. This can be due to infrastructural issues or any other serious issue that is beyond the application code. For such scenarios, we use **uptime monitoring**. In our next section, we shall explore how we can use uptime monitoring to ensure better **Service-Level Agreements (SLAs)** for our Django application.

Integrating uptime monitoring

Would you use an application that frequently goes down? No. When you create an application and users are using it, you need to make sure your service has maximum availability. The uptime SLA is crucial for every service to establish trust among users. In this section, we shall learn how we can monitor the uptime of our Django application by adding a health check endpoint to our automated monitoring system.

Adding a health check endpoint

django-health-check (https://github.com/revsys/django-health-check) is a third-party package that can be easily added to any Django project to create an endpoint that can be monitored for uptime. Let us learn how to integrate django-health-check into the Django project. To do so, follow these steps:

1. Install django-health-check in our Django project by using the following code in our terminal:

    ```
    pip install django-health-check
    ```

2. Now add health_check to Django INSTALLED_APPS in the settings.py file:

    ```
    INSTALLED_APPS = [
        # …
        'health_check',
        'health_check.db',          # Checks database connection
        'health_check.cache',       # Checks cache backend
    ]
    ```

3. Add the django-health-check endpoint to the project in the main urls.py file:

    ```
    urlpatterns = [
        # ...
        path(r'^ht/', include('health_check.urls')),
    ]
    ```

4. Now, when we hit the /ht/ endpoint in the browser, we will see the health check of our system. In *Figure 14.5*, we have a screenshot of a live system with multiple systems integrated with django-health-check:

System status

Service		Status	Time Taken
✓	Cache backend: default	working	0.0261 seconds
✓	CeleryPingHealthCheck	working	1.0273 seconds
✓	DatabaseBackend	working	0.0225 seconds
✓	DefaultFileStorageHealthCheck	working	0.0163 seconds
✓	DiskUsage	working	0.0001 seconds
✓	MemoryUsage	working	0.0002 seconds
✓	MigrationsHealthCheck	working	0.0568 seconds
✓	RedisHealthCheck	working	0.0125 seconds

Figure 14.5: django-health-check output

Now we have a health check endpoint that can be opened in our browser to see whether our service is up and running whenever we want. Let us add automation to regularly monitor our Django application. There are numerous third-party tools, such as Pingdom, BetterStack, FreshPing, New Relic, and so on, that one can use. We shall use **BetterStack** (https://betterstack.com/) in this chapter.

Using BetterStack for uptime monitoring

In this section, we shall learn how to use the BetterStack platform for website uptime monitoring. BetterStack has a generous free tier that we will use to create website monitors by following these steps:

1. Create a new account in BetterStack (https://betterstack.com/).

2. Once we create a new account and log in to our dashboard, click on the **Create monitor** button:

Figure 14.6: Create a new uptime monitor using BetterStack

3. Now create the monitor by adding the django-health-check endpoint in the **URL to monitor** field. The django-health-check endpoint will send a response status code of 200 whenever the health check is successful but will respond with a 500 status code if any of the health checks fail.

4. We will leverage the status code and configure the BetterStack monitor to check if the response code is 200. If so, it will be a success. Otherwise, the health check will be marked as a failure. In *Figure 14.7*, we are using a *mock postman* URL (`https://<mockid>.mock.pstmn.io/ht`) as an example and have configured **Alert us when** – **URL returns HTTP status other than** – 200.

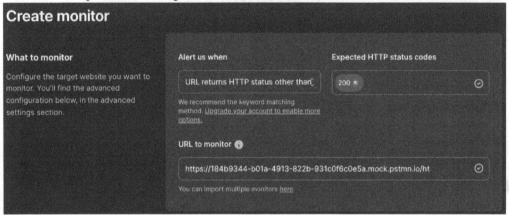

Figure 14.7: Configure the health check URL that we want to monitor

Now, whenever our server has a health check issue, the endpoint will respond with 500, and this will result in the uptime monitor failing.

5. The BetterStack official documentation has the latest steps to integrate Slack with BetterStack. Please follow the official documentation at `https://betterstack.com/docs/uptime/slack/`.

    Once we have our Slack integration set up with BetterStack, BetterStack will send a Slack notification whenever our site goes down.

6. BetterStack provides a few other integrations that can be easily integrated. Once we configure Slack with BetterStack, we can now see any incident notification on our Slack channel and other communication mediums.

Figure 14.8: Slack message when the uptime monitor fails

We have successfully configured our Uptime monitor to check for any downtime for our website. Whenever there is downtime on our website, the engineers will get informed about the incident first and they can then start looking into the issue.

On-call process

Every team should have an on-call engineer who will be available to acknowledge any incident triggered by the uptime monitoring tool or other APM alerts. It is important to set up a good on-call schedule that supports a better SLA for production services.

In this book, we have not shown how to set up an on-call process as it is out of scope, but you can use different tools that can help in managing the on-call schedule, for example, SquadCast, PagerDuty, and so on.

The official documentation for incident management tools is straightforward, and I would recommend you explore these tools for in-depth understanding. Here are a few resources for further reading: `https://www.pagerduty.com/`, `https://www.squadcast.com/`, and `https://www.atlassian.com/software/opsgenie`.

We have learned how to monitor code-level errors and any service downtime. As users start to use the Django application, we will hit different scales and performance bottlenecks. **Application Performance Monitoring (APM)** tools are used to monitor any performance bottleneck that occurs in production. In the next section, we shall learn how to integrate APM tools into our Django application.

Integrating APM tools

Suppose we develop, test, and optimize our application on our local development setup. Now, when we deploy our changes to production, everything is working fine, but suddenly, after a week, users complain about slow response times and panic mode starts. What do we do now? How do we debug the reason for degraded performance in production? To solve this problem, we use APM tools such as New Relic.

APM tools help us identify all performance bottlenecks and root causes of them. Degraded performance can be due to a badly written DB query, badly written application code, or maybe a higher number of user requests. Whatever the reason, APM tools can help us identify the issues and guide us to the right root cause with less guesswork and more data-driven conclusions.

There are different APM tools available, such as New Relic, DataDog, Chronosphere, Splunk, and so on. We shall learn about New Relic in this book, but you can easily follow the official documentation of other tools to integrate the APM tool.

New Relic (`https://newrelic.com/`) has a generous free tier that provides 100 GB of free data ingestion per month. The New Relic free tier is good enough for any new application to monitor the initial set of customers.

> **Learn more about New Relic**
>
> New Relic has many features and it can be overwhelming for anyone new to get started with. It is also difficult to cover all the details of New Relic in one section. So, we shall limit our scope only to the basic features of New Relic and let you explore more as you become more familiar with the platform.
>
> New Relic also has a free-of-cost course and certification. For more details, visit the *New Relic University* website at `https://learn.newrelic.com/`.

Now let us start with the integration of New Relic into the Django project.

Integrating New Relic into the Django project

We shall use New Relic as an APM tool and also as a log monitoring tool in our project. First, let us learn how to integrate New Relic into the Django project for APM.

Let us see an overly simplified explanation of how New Relic works – New Relic will be placed between the user and the Django application as shown in *Figure 14.9*:

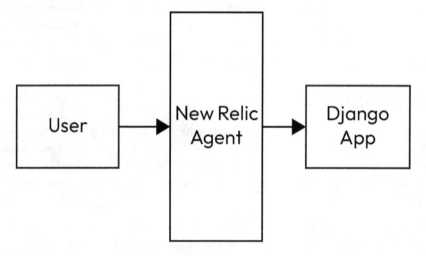

Figure 14.9: The New Relic agent added to the Django application

Let us create an account in New Relic and start the APM integration:

1. Click on **Add Data** from the side navigation bar and select **Application monitoring | Python** as shown in *Figure 14.10*:

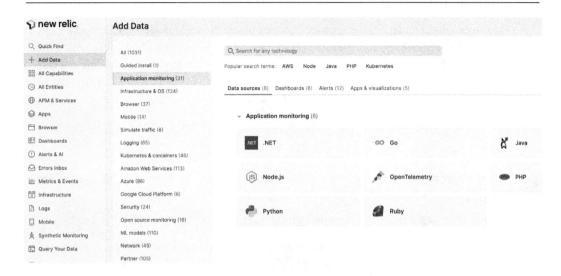

Figure 14.10: Showing how to integrate New Relic into Django

Once we select the **Python** agent, New Relic takes us to a guided wizard that helps us to directly integrate New Relic into our Django application by following the next steps.

2. As shown in *Figure 14.11*, we need to give a name to our Django application, we have used `Django Demo App`.

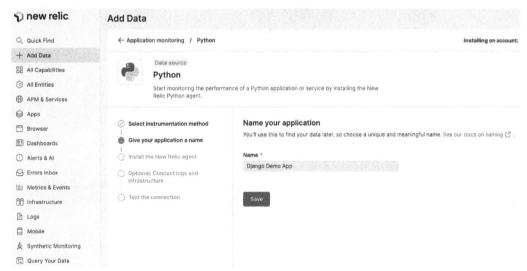

Figure 14.11: New Relic wizard for Python integration

3. Next, follow the steps mentioned in the wizard. Install the Python APM agent:

```
pip install newrelic
```

Save the `newrelic.ini` file in the root directory where the `manage.py` file is present. (The content of the `newrelic.ini` file can be downloaded from the wizard. We are not showing the content of the file since it would be copied and pasted, and also it is too big of a file). You can take a look at `newrelic.ini` file at `https://github.com/PacktPublishing/Django-in-Production/blob/main/Chapter14/myblog/backend/newrelic.ini`

4. Now validate the configuration by using the following command:

```
newrelic-admin validate-config newrelic.ini
```

5. Once the configuration is validated, we need to run the plugin in the New Relic agent in our Django app, by running the following command:

```
newrelic-admin run-program python manage.py runserver
```

The preceding command is to test the setup on our local machine. We are using a `docker` and `gunicorn` setup to run our Django application in production. We will need to update our `docker-compose.yaml` file to integrate New Relic into the Django application. If you are using some other setup, such as systemd or any other setup, then you will need to add the `newrelic-admin run-program` command before the `gunicorn` command as we have shown here:

```
services:
  backend:
    build:
      context: ./backend
      dockerfile: Dockerfile.prod
      command: newrelic-admin run-program gunicorn backend.
  prod_wsgi:application --workers 4 --bind 0.0.0.0:8000
  --max-requests=512 --max-requests-jitter=64 --reload
```

Now as we deploy the changes to production, we will start receiving data for our web transactions on the New Relic dashboard.

We have successfully integrated our Django application with New Relic to capture performance metrics.

> **Important note**
> Since we need real-world data and a considerable amount of traffic for New Relic to show its capabilities, I will be using one of my side projects as a reference. I will be hiding any sensitive data from the screenshots due to security concerns.

Exploring the New Relic dashboard

Let us now check out how our New Relic APM dashboard looks when we have data sent from the Django server. In *Figure 14.12*, we can see three different sections highlighting the three metrics that are important for performance monitoring:

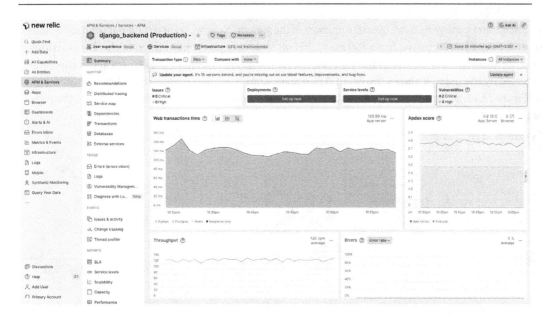

Figure 14.12: New Relic dashboard showing performance metrics for Django app

Let us look into these three metrics in detail:

- **Web transaction time:** This is the average response time of all APIs in our Django application. The web transaction time is calculated as the request-response time taken by the Django application and not the time required for the request to travel from the user to the Django server. Though the web transaction time is the average time for all the requests served by the application, having a lower web transaction time shows good application code and well-thought-through architecture. One can drill down further and analyze individual request-response transaction times by going into the **Transactions** section.

- **Throughput of Django app**: Throughput shows the total number of requests served by the Django application. This number is represented by **Requests per Minute (RPM)**. As the user base grows, this number will increase. Keeping a close eye on RPM spikes is important to the stability of the system.

- **Apdex score:** The **Application Performance Index (Apdex)** is an industry standard to measure users' satisfaction with the response time of web applications and services. Keeping a stable Apdex score should be the goal of an engineering team. It is a complicated calculation, and we shall not go into the details of how to calculate it. You can read more at https://docs. newrelic.com/docs/apm/new-relic-apm/apdex/apdex-measure-user-satisfaction/.

The **Summary** page gives us an overall summary of the system. Now, we want to find out why our system response time has increased from *100 ms* to *300 ms* suddenly, so we need to go to the **Transactions** tab. The **Transactions** tab shows the list of all the APIs that are present in the Django application and accessed by the users. It shows the list of API transactions from maximum time taken to minimum. In *Figure 14.13*, we can see the **Transactions** tab for the Django application (some parts of the page are blurred for security reasons):

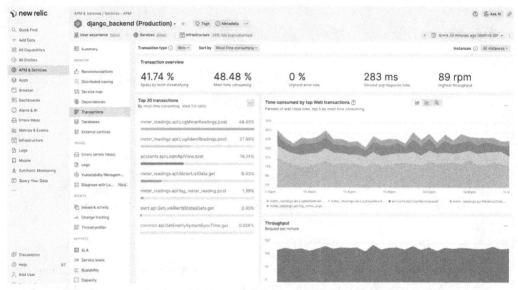

Figure 14.13: Transactions tab for the Django application

Using the **Transactions** tab, we can find all the slow endpoints or any endpoint that has more throughput, and so on. Suppose we found an API that is taking *85 ms* and that is considered to be slow as per the Django application. By selecting a web transaction, we can check more details of the application. In *Figure 14.14*, we can see the details of a particular transaction and analyze which part of the application is taking the most time (some parts of the page are blurred for security reasons):

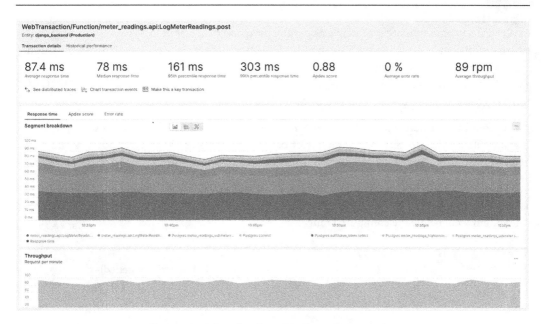

Figure 14.14: Details of a particular web transaction

Figure 14.14 shows in detail if the Postgres query takes time or if any other application code is taking time. Such detailed analysis can help us get into the root cause of any issue with a proper data-driven approach.

Creating New Relic alert conditions

With all this data present in New Relic, we can create automated alerts that will alert the team whenever there is a deviation in performance. New Relic provides an interface to create custom alerts and policies.

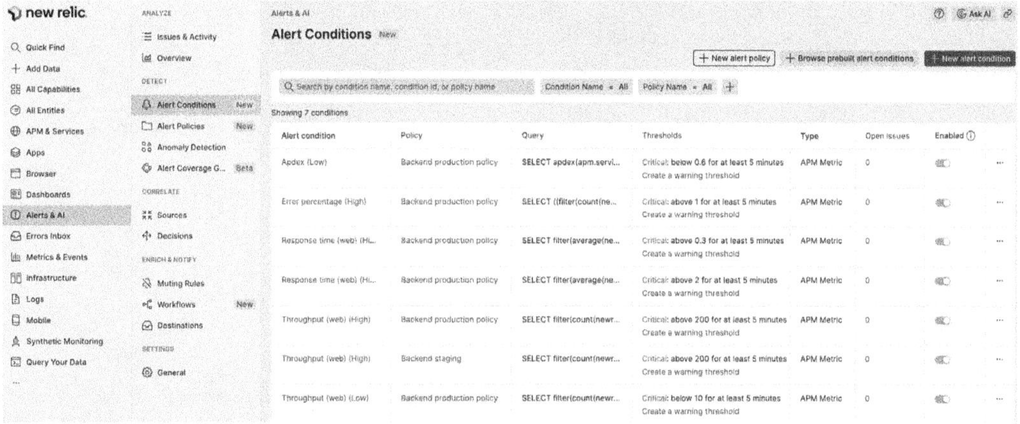

Figure 14.15: New Relic custom alert conditions and policies

New Relic alert conditions are extendable as per the need and data captured by New Relic. Standard alert conditions can be set for an increase in web transaction time, spikes in throughput, degradation in Apdex score, and so on. For more details, check the official documentation: `https://docs.newrelic.com/docs/alerts-applied-intelligence/new-relic-alerts/alert-conditions/create-alert-conditions/`.

Once we configure an alert, we should also integrate Slack with New Relic. The official documentation shows the detailed steps for Slack integration (`https://docs.newrelic.com/docs/alerts-applied-intelligence/notifications/notification-integrations/#slack`).

Now, whenever New Relic detects an anomaly for which we have configured an alert condition, it will raise an alert. In *Figure 14.16*, we can see a Slack message sent by New Relic because the throughput of the system went below the configured range:

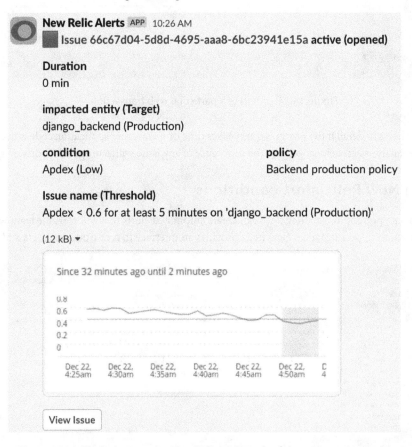

Figure 14.16: Slack alert sent by New Relic on breach of throughput threshold

Creating the right New Relic alerts for the web application can ensure better SLA monitoring and deploying the right amount of servers to production, along with identifying any hidden performance bottlenecks.

Monitoring AWS EC2 instances with New Relic

In *Chapter 13*, we learned how to use AWS Beanstalk to deploy our Django application. Our Django application is deployed using AWS Beanstalk, so we need to integrate our New Relic infrastructure agent on AWS Beanstalk. New Relic has a straightforward guide to installing the New Relic infrastructure agent in all Beanstalk-managed EC2 instances. The integration steps are beyond the scope of this book. Please follow the guide mentioned here: `https://docs.newrelic.com/docs/ infrastructure/install-infrastructure-agent/config-management-tools/ configure-infrastructure-agent-aws-elastic-beanstalk/`.

With the New Relic infrastructure agent, we can monitor the RAM, CPU usage, disk space, and additional data for the EC2 host machine. Now that we have the infrastructure agent configured for our EC2 instances, let us start sending all the log files that are present on the host machine.

Sending logs from Django to New Relic

In *Chapter 6*, we learned how to set up logging in Django. We configured the path of log files in `settings.py` in the `filename` section of the `LOGGING` config.

Now we will tell the New Relic infrastructure agent to watch the file content of `path/to/django_ logs.log` and send the logs to New Relic:

```
LOGGING = {
    "version": 1,
    "disable_existing_loggers": False,
    "formatters": {"verbose": {"format": "%(asctime)s %(process)d
%(thread)d %(message)s"}},
    "loggers": {
        "django_default": {
            "handlers": ["django_file"],
            "level": "INFO",
        },
    },
    "handlers": {
        "django_file": {
            "class": " logging.handlers.RotatingFileHandler",
            "filename": "path/to/django_logs.log",
            "maxBytes": 1024 * 1024 * 10,   # 10MB
            "backupCount": 10,
            "formatter": "verbose"
        },
    },
}
```

We have already discussed how the logging configuration works in Django, in *Chapter 6*.

> **Important note**
>
> The New Relic infrastructure agent must be installed on the host machine to send logs from the host system to the New Relic system. If you are not using any other setup, please follow the documentation provided by New Relic to install the infrastructure agent on the host machine before following the next steps.

To set up logging in our Django project and forward the logs to New Relic, we need to perform the following steps:

1. Create a `logging.yml` config file that will tell the New Relic infrastructure agent where to pick the appropriate log files and send the data to the New Relic server. The `logging.yml` file should have the following content inside it:

    ```
    logs:
        - name: "django_log"
          file: /<log file path>/*.log
        - name: "nginx_log"
          file: /<nginx log file path/*.log
        - name: "etc_log" # any other log we want to send
          file: /path/*.log
    ```

    name key value can be anything we decide, and `file` should give the absolute path of the folder where log files are generated.

2. Once we have created the `logging.yml` config file, we need to save the `logging.yml` file in `/etc/newrelic-infra/logging.d/logging.yml` on the host machine – in our case, the EC2 instance where Beanstalk is running our Django application. Save the `logging.yml` file in the `.ebextensions` folder in the project root.

3. We can set up logging for the Django app by using `.ebextensions`. Create a `02_newrelic_logging.config` file with the following code snippet:

    ```
    container_commands:
        01_copy_logging_config :
            command: 'cp .ebextensions/logging.yml /etc/newrelic-
    infra/logging.d/'
    ```

4. Now, with the fresh deployment of our Django application, all the logs generated by the Django application will be accessible in New Relic.

Now, if we open the **Logs** tab in our New Relic dashboard, we can see the logs collected by the Django application and sent to New Relic. *Figure 14.17* shows a list of logs captured and sent to New Relic. New Relic provides Lucene search to scan through all the logs (some parts of the page are blurred for security reasons).

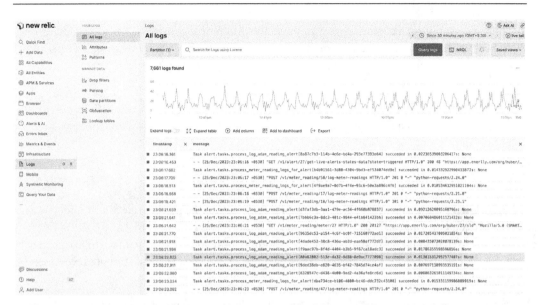

Figure 14.17: An overview of logs captured by Django and sent to New Relic

If we want to check the details of particular log data, we can open the selected log and check the details. In *Figure 14.18*, we can see a detailed example of a log (some parts of the page are blurred for security reasons):

Figure 14.18: Detail log in New Relic dashboard

New Relic also supports a custom parser to parse the different log formats. For more details on how to best work with infrastructure agent logs, check the official documentation: `https://docs.newrelic.com/docs/logs/forward-logs/forward-your-logs-using-infrastructure-agent/`.

New Relic is capable of storing multiple sources of data on its platform. To work with hundreds of GBs of data, one needs to query the data. For such purposes, New Relic has a query language that is similar to SQL.

Working with metrics and events using NRQL

New Relic Query Language (**NRQL**) is a powerful tool that can be used to query and understand all types of data. One can easily learn about and start using NRQL to analyze data on the platform and also create charts and dashboards. In the beginning, it might be overwhelming to learn a new query language to use monitoring, but from my personal experience, I would strongly recommend learning NRQL.

The official documentation is easy to follow and it will hardly take a couple of hours to learn the syntax of NRQL. Please follow the documentation at `https://docs.newrelic.com/docs/query-your-data/nrql-new-relic-query-language/get-started/introduction-nrql-new-relics-query-language/`.

When we go to the **Metrics & Events** tab, we can see all the events captured in New Relic. By selecting **Event type**, we can start exploring the data. In *Figure 14.19*, we have selected the **Transaction** events and now we can explore the data using NRQL.

Figure 14.19: Exploring Metrics & Events

Now that we have learned how to work with multiple tools, such as APM, error monitoring tools, uptime monitoring, and so on, it can become daunting to have so many tools to work with. Modern engineering is more about optimizing and using the right tool for the job. In our next section, we shall learn how to optimize our workflow with the Slack messaging tool.

Integrating messaging tools using Slack

In today's software development world, messaging tools such as Slack/Teams are an integral part of the development cycle. In this section, we shall take Slack (`https://slack.com/`) as an example and show how software developers can take advantage of such communication tools to improve their productivity and enhance monitoring.

In the *Integrating Rollbar with Slack* and *Creating New Relic alert conditions* sections, we showed how we can integrate Slack into error monitoring tools, uptime monitoring tools, and APM tools. This way, developers do not have to go to all applications individually to keep track of whether an alert is triggered or not. Rather, whenever there is an alert, the tool will automatically message on Slack and developers access the details from Slack directly.

Slack provides a rich message interface, so users can not only respond to messages but also take action without leaving Slack. For example, when Rollbar notifies us about an error, we can directly tag the person about the error and have a conversation. We can also assign the issue to the developer without going to Slack, as shown here:

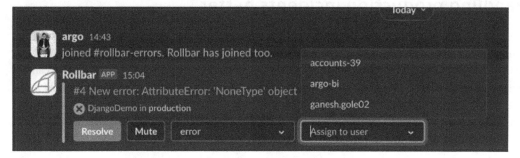

Figure 14.20: Assigning Rollbar issues to users directly from Slack

We can also click the **Resolve** button to mark an issue as resolved directly from Slack.

Similarly, whenever there is an incident triggered by BetterStack, we can acknowledge the incident from Slack. These Slack message actions help developers to have all the tools and important actions in one place, also making sure the visibility of any critical incident is not missed by engineering leaders.

Today, almost every third-party tool has Slack integration because it helps consolidate all information in one place and makes sure Slack notifications are not missed. Whether it is Jira, GitHub, or Google Calendar, all these Slack applications help make sure developers are not missing out on important notifications. It would not be wrong to say that I spend 20% of my workday on Slack and almost 90% of my notifications are managed via Slack.

Here are some more official resources by Slack on how Slack can improve the overall software development cycle and developer productivity:

- `https://slack.com/intl/en-in/resources/using-slack/speed-up-software-development-with-slack`

- `https://slack.com/intl/en-in/resources/why-use-slack/the-value-of-slack-for-software-developers`

- `https://slack.com/intl/en-in/resources/why-use-slack/software-development-teams-and-slack`

- `https://slack.com/intl/en-in/resources/why-use-slack/how-it-works-slack-for-software-development`

- `https://slack.com/intl/en-in/resources/why-use-slack/how-slacks-own-developers-use-slack`

We integrate APM tools and uptime monitors into our system so that we are automatically notified about a system degradation rather than customers complaining about our service. Whenever a system goes down or has degraded performance, we call it an incident. Production incidents can be of different severity levels, depending upon the impact. In the next section, we shall focus on how different teams can work when the production system goes down.

Handling production incidents better

Every on-call engineer's nightmare is getting a call in the middle of the night about production systems being down. Production incidents are common in every company, be it a two-engineer start-up or a 20,000-engineer big tech organization such as Google. An incident is defined as an event that causes disruption or degraded performance to the end user using a service. In the *Integration APM tools* section, we learned how to use APM tools to identify degraded performance and how to use uptime monitoring tools to identify any downtime and alert stakeholders. Now let us learn how to work during an incident and how to manage things better.

The first job of an on-call engineer is to make sure they don't panic. Contrary to common opinion, I have observed that engineers who do not take incidents as a "do or die situation" can handle incidents much better. It is very difficult to have a generic approach to solving production incidents since the reasons for them will be different every time. But having a **Standard Operating Procedure** (**SOP**) for incident management is important.

Let me list down a few SOPs I have learned while working with different companies across industries:

- **Communicate first**: It is important to communicate/acknowledge any incident first before jumping into solving it. The communication should be first done in the internal messaging channel and then with external users if needed.

- **Identify the issue before taking any action**: The most important step during an incident is to identify the issue that is causing an incident. If we have a high response time for the API, we need to go to our APM tool to identify what is causing the higher response time. Is it due to the increase in throughput, due to a bad query, or because of some downstream dependency having a slower response time? Depending upon the root cause, we should take action.

- **Involve stakeholders**: During an incident, always try to do the initial analysis first, but if the root cause is still unclear, it is important to involve the other stakeholders.

- **Set up a war room**: Have all the stakeholders in one meeting, work synchronously with all the people involved, and have verbose communication with each member and each role.

- **Revert before fixing**: Revert first, fix later should always be the approach. Whenever you find an issue caused by a recent code change, the first intuition of most on-call engineers is to fix the bug, but that should be the second step. The first step for an on-call engineer should be to revert the code changes to stabilize the production system first and later work on fixing the issue.

- **Mitigation and monitoring**: Once the issue is mitigated, we should keep an eye on the improvement of system health and make sure we closely monitor the system for the next couple of hours.

- **Creating Root Cause Analysis (RCA) and action items**: Once the production systems are stable, it is important to make sure the team performs a blameless RCA and creates action items to avoid such incidents in the future. An incident can be closed only after the RCA is approved and action items from the incident are implemented.

We have learned about an SOP for incident management, but every team and company can have different procedures to handle incidents. Some tools can help in enforcing such procedures, and tools such as Squadcast (`https://www.squadcast.com/`), and Opsgenie (`https://www.atlassian.com/software/opsgenie`) can be used to streamline incident management.

In our next section, we shall learn about how to create a **blameless RCA culture** in the team for incidents. It is important to create a healthy blameless culture for incident reviews so that engineers feel comfortable and show ownership and empathy toward fellow engineers.

Blameless RCA for incidents

RCA can be defined as a systematic method to uncover the fundamental cause of an incident through a series of "why?" questions until no further diagnostic information can be extracted. The most important part of conducting an RCA is being *blameless*. In this section, we shall learn how to create a blameless RCA for incidents that would help in creating a strong and healthy engineering culture in the team during RCAs.

Let us learn about a few important pointers that can help us create a blameless RCA:

- **Focus on what went well during the incident**: Highlighting the positive points that went well during the incident gives more confidence to the team. Thank the on-call engineers who responded to the incident and did the firefighting.

- **Focus on the future**: Do not focus on what could've happened, should've happened, or any past events. Rather, focus on what actions we can take to improve for next time or avoid it happening again.

- **Focus on work and not on people**: Always focus on why something failed, rather than who was working on it.

- **Show empathy and avoid ego**: No one intentionally adds a bug or creates a faulty design, so being empathetic toward the other engineer is important.

- **Accept mistakes**: We mustn't take feedback personally and try to defend everything. Feedback is always about the work and not about the person, so accept the mistake and move on to the next step of how to improve.

Every company can have its own culture on how to implement a blameless RCA. Here are a few blameless RCAs that different companies follow in the industry:

- Gitlab: `https://handbook.gitlab.com/handbook/customer-success/professional-services-engineering/workflows/internal/root-cause-analysis/`

- Atlassian: `https://www.atlassian.com/incident-management/postmortem/blameless`

- PagerDuty: `https://postmortems.pagerduty.com/culture/blameless/`

Once we have a good culture set for blameless RCA, we can create an RCA template that can be followed by teams. We shall not get into the details of an RCA template since every company would prefer to have its own implementation. Here are a few examples of RCA templates that are widely popular across the industry:

- Gitlab template: `https://gitlab.com/gitlab-org/gitlab/-/blob/master/.gitlab/issue_templates/rca.md`

- Atlassian template: `https://www.atlassian.com/incident-management/postmortem/templates#incident-summary`

- Google template: `https://sre.google/sre-book/example-postmortem/`

For more templates, check out the repository at `https://github.com/dastergon/postmortem-templates`. As teams mature and handle incidents, these templates will evolve and be more relevant to the company.

In this section, I tried to focus on the importance of blameless RCA since this sets the right engineering culture and, most of the time, gets overlooked in early-stage start-ups if there are very few engineers. A good engineering leader always focuses on the action items for the present and how to improve systems in the future rather than what could have been done better in the past. A blameless RCA would encourage engineers to take ownership of the incident and work on action items to improve the system.

Summary

In this chapter, we have learned how to work with different tools to monitor and improve our application. APM tools such as New Relic help in application performance monitoring, which is critical for any application. We learned how to integrate New Relic into our Django application. Getting 5xx errors in production is every developer's nightmare, and error monitoring tools such as Rollbar and Sentry help in capturing and tracking bugs so that developers can fix them easily. We have also learned in this chapter how to integrate uptime monitoring tools that continuously monitor our application and inform us immediately if there is downtime.

It is important to know that in today's application development process, writing code is the first part, and using the right tools for maintenance and communication is crucial. With so many third-party tools, it can be overwhelming for developers and other engineering leaders to track every tool, hence unifying all the communication and alerts in one place, such as Slack/Teams, is crucial. We have learned how we can integrate Slack/Teams into our development process. One needs to accept that production incidents are a reality. Every team will face a production incident and developers must be well prepared to handle such incidents with appropriate processes. Once a production incident occurs, it is important to reflect on the issue and create a blameless RCA that can avoid such occurrences in the future. In this chapter, we have learned how to manage incidents better and create RCA that can help us improve systems further.

We have now come to the end of our book – *Django in Production*. We have learned how to create a web application using Django, Django REST Framework, and Celery. We explored AWS and how we can deploy our application to production. We have tried to cover all the important topics that a software developer should be aware of while building a web application using Django. A few of the advanced concepts are out of the scope of the book and we have redirected you to learn about them from official documentation. In this book, we have discussed a lot of best practices and practical use cases that I as a developer have learned while working in the software industry.

I hope you have enjoyed all the topics mentioned in the book. Good luck with all your Django adventures!

Index

A

packtpub.com

Subscribe to our online digital library for full access to over 7,000 books and videos, as well as industry leading tools to help you plan your personal development and advance your career. For more information, please visit our website.

Why subscribe?

- Spend less time learning and more time coding with practical eBooks and Videos from over 4,000 industry professionals

- Improve your learning with Skill Plans built especially for you

- Get a free eBook or video every month

- Fully searchable for easy access to vital information

- Copy and paste, print, and bookmark content

Did you know that Packt offers eBook versions of every book published, with PDF and ePub files available? You can upgrade to the eBook version at packtpub.com and as a print book customer, you are entitled to a discount on the eBook copy. Get in touch with us at customercare@packtpub.com for more details.

At www.packtpub.com, you can also read a collection of free technical articles, sign up for a range of free newsletters, and receive exclusive discounts and offers on Packt books and eBooks.

Other Books You May Enjoy

If you enjoyed this book, you may be interested in these other books by Packt:

Django 5 By Example

Antonio Melé

ISBN: 978-1-80512-545-7

- Implement different modules of the Django framework to solve specific problems
- Integrate third-party Django applications into your project
- Build asynchronous (ASGI) applications with Django
- Set up a production environment for your projects
- Create complex web applications easily to solve real use cases
- Implement advanced functionalities, such as full-text search engines, user activity streams, payment gateways, and recommendation engines
- Add real-time features to your apps with Django Channels and WebSockets

Mastering Flask Web Development

Sherwin John C. Tragura

ISBN: 978-1-83763-322-7

- Prepare, set up, and configure development environments for both API and web applications
- Manage request and response data with Flask's request and response
- Integrate Flask applications with NoSQL databases like HBase/Hadoop, Cassandra, Mongo, and Neo4J to solve big data problems
- Apply various ORM and ODM to build the model and repository layer of the applications
- Integrate with OpenAPI, Circuit Breaker, ZooKeeper, OpenTracing and others to build scalable API applications
- Utilize Flask middleware to provide CRUD transactions for Flutter-based mobile applications
-

Packt is searching for authors like you

If you're interested in becoming an author for Packt, please visit authors.packtpub.com and apply today. We have worked with thousands of developers and tech professionals, just like you, to help them share their insight with the global tech community. You can make a general application, apply for a specific hot topic that we are recruiting an author for, or submit your own idea.

Share Your Thoughts

Now you've finished *Django in Production*, we'd love to hear your thoughts! Scan the QR code below to go straight to the Amazon review page for this book and share your feedback or leave a review on the site that you purchased it from.

https://packt.link/r/1804610488

Your review is important to us and the tech community and will help us make sure we're delivering excellent quality content.

Download a free PDF copy of this book

Thanks for purchasing this book!

Do you like to read on the go but are unable to carry your print books everywhere?

Is your eBook purchase not compatible with the device of your choice?

Don't worry, now with every Packt book you get a DRM-free PDF version of that book at no cost.

Read anywhere, any place, on any device. Search, copy, and paste code from your favorite technical books directly into your application.

The perks don't stop there, you can get exclusive access to discounts, newsletters, and great free content in your inbox daily

Follow these simple steps to get the benefits:

1. Scan the QR code or visit the link below

https://packt.link/free-ebook/9781804610480

2. Submit your proof of purchase
3. That's it! We'll send your free PDF and other benefits to your email directly

www.ingramcontent.com/pod-product-compliance
Lightning Source LLC
Chambersburg PA
CBHW080618060326
40690CB00021B/4741